Nonlinear Systems and Complexity

For further volumes:
http://www.springer.com/series/11433

Luis Vázquez • Salvador Jiménez

Newtonian Nonlinear Dynamics for Complex Linear and Optimization Problems

Luis Vázquez
Departamento de Matemática Aplicada
Facultad de Informática
Universidad Complutense de Madrid
and
Real Academia de Ciencias Exactas, Físicas
  y Naturales
Spain

Salvador Jiménez
Departamento de Matemática Aplicada a las
  TT.II.
E.T.S.I. Telecomunicación
Universidad Politécnica de Madrid
Spain

ISBN 978-1-4939-0017-6   ISBN 978-1-4614-5912-5 (eBook)
DOI 10.1007/978-1-4614-5912-5
Springer New York Heidelberg Dordrecht London

Springer is part of Springer Science+Business Media (www.springer.com)

To Luciana and José Luis with whom
I enjoyed many fierce pinball matches
in El Escorial and Forte dei Marmi,

Luis Vázquez

To my beloved ones,

Salvador Jiménez

Preface

In a *Pinball Machine*, the player tries to score points by manipulating a metal ball on a playing field inside a glass covered case. The objectives of the game are to score as many points as possible, to earn free games and to maximize the time spent playing by earning extra balls and keeping balls in play as long as possible. Apart from the new challenging features, the good old pinball playing field is essentially a planar surface inclined upwards from 3 to 7°, away from the player, and include multiple targets and scoring objectives. The ball is put into play by the use of the *plunger* which propels upwards the ball. Once the ball is in play, it tends to move downwards towards the player, although the ball can move in any direction, sometimes unpredictably, as the result of contact with objects on the playing field or by the player actions. To return the ball to the upper part of the playing field, the player makes use mainly of one or more *flippers*. The game ends whenever the ball crosses downwards the "flippers barrier" [26].

The *Pinball Machine* provides a simple mechanical example of the linear optimization problem, basically in a surface embedded in the three-dimensional space. In all pinball games, the play with every ball finishes when that ball reaches the minimum gravitational potential energy immediately after the flippers barrier. On the other hand, the playing field where the ball moves is, effectively, a convex planar region. The duration of the ball motion is always finite, even considering the human interaction. This fact indicates that the minimum of the *Objective Function* (in this case, the Potential Gravitational Energy) is always attained by the motion of the ball. This example suggests us to associate the solutions of some optimization problems to the motion of Newtonian particles. At the same time, this example is a bridge that allows us to construct algorithms for linear/nonlinear optimization problems and unconstrained extrema by applying to them the numerical algorithms used to simulate the equations of motion for a Newtonian particle. These are the motivation and the objective of this monograph. The framework of this monograph wants to be constructive: we want to present some methods and their features that show how Newton's equation for the motion of one particle in classical mechanics combined with finite difference methods can create a mechanical scenario within which we may solve some basic, though complex, problems. We, thus, apply these ideas to solve linear systems and eigenvector problems, as well as programming, both linear and nonlinear, in different dimensions, in the spirit of the suggestive books by Mordecai Avriel [3] and John T. Betts [5]. For this latter case, the goal of the monograph is to show a breakthrough analysis method of optimization by combining the features of the motion of a Newtonian classical particle and finite difference numerical algorithms associated with the equation of motion. Many challenging questions remain open, but we think that a new, fresh and feasible approach to solve them is shown.

This unified numerical and mechanical approach is new, to the best of our knowledge, and we believe that our view represents a simple but useful tool not yet fully exploited.

This monograph is intended for a broad public: undergraduate and graduate students or researchers who are confronted in their work with linear systems and eigenvalue or optimization problems and who are open to new perspectives in the way these problems can be addressed. To help the reader to explore these ideas, we propose a list of related exercices at the end of each chapter.

The basic mechanical equations and assumptions are presented in Chap. 1: we review the basic laws for the motion of a particle under Newton's second law, in one and several dimensions, with and without dissipation. Different cases, depending on the acting potential, are presented. We also present two numerical schemes to simulate the corresponding equations of the motion. All this material should be thoroughly used in the sequel as basic building blocks with which to construct methods to solve the proposed problems, ranging from linear algebra to nonlinear programming.

In Chap. 2 we propose a new iterative approach to solve systems of linear equations. The new strategy integrates the algebraic basis of the problem with elements from classical mechanics and the finite difference method. The approach defines two families of convergent iterative methods. Each family is characterized by a linear differential equation, and the methods are obtained from a suitable finite difference scheme to integrate the associated differential equation. These methods are general and depend on neither the matrix dimension nor the matrix structure. We present the basic features of each method. As a consequence, we also present a general method to determine whether a given square matrix is singular or not.

In Chap. 3 we apply the previously developed methods to several examples. We compare these with other similar characteristics, such as Jacobi, Gauss–Seidel, and Steepest Descent Methods and discuss several aspects about choosing the parameter values for the numerical methods.

In Chap. 4 we consider the computation of eigenvectors and eigenvalues of matrices. For a general square matrix, not necessarily symmetric, we construct a family of dynamical systems whose state converges to eigenvectors which correspond to eigenvalues with smallest and biggest real part. We further analyse the convergence and perform several numerical tests. Besides, we extend the application of the method to the effective computation of all eigenvalues with intermediate real part. Some examples and comparisons with the Power Methods are presented in

Chap. 5. We design some ways to enhance the linear convergence of the method, combining it with two different quadratic methods.

In Chaps. 6 and 7 we apply our ideas to solve the so-called programming problems. Chapter 6 is devoted to the classical linear programming problem. We propose a new iterative process to approach the solution of the Primal Problem associated with the linear programming problem: $\max Z = C^{\mathrm{T}} \vec{x}$, with some linear constraints. The method is based on translating the problem to the motion of a Newtonian particle in a constant force field. The optimization of the objective function is related to the search for the minimum of the particle's potential energy. Several solution strategies which depend on the number of dimensions are developed and also illustrated through different examples.

The monograph comes to an end in Chap. 7, which is devoted to the classical quadratic programming: we extend our previous method to the case of optimizing a quadratic objective function with linear constraints as well as to the case of a linear function with quadratic constraints. The method can also be extended to the general case of a nonlinear objective function with linear constraints.

This work has been partially supported by several different research grants. Particularly, we thank the project "The Sciences of Complexity" (ZiF, Bielefeld Universität, Germany) and the hospitality of the Centro de Ciências Matemáticas (Universidade da Madeira, Portugal) where part of this work was done. We wish also to thank, very specially, our friends and colleagues Profs. Pedro J. Pascual and Carlos Aguirre for their comments and suggestions. Also, we are indebted to our friend Prof. Robin Banerjee for his support and enlightening comments.

Madrid, Spain   Luis Vázquez
Madrid, Spain   Salvador Jiménez

Contents

Chapter 1
Elements of Newtonian Mechanics

1.1 Introduction

Within the framework of Newtonian Mechanics, the dynamics of the motion of a single particle is related to the following properties: (1) *the existence of either conservation or variation laws for some quantities according to certain relations*, (2) *the existence of symmetries* and (3) *the stability character of the equilibrium points of the system*. Only in a few cases the analytical solution of the equations of motion is known and, for this reason, numerical simulations are usually necessary. In order to get accurate numerical solutions, it is desirable to use numerical schemes that show a discrete analogy of the properties listed above. If any of these properties is not satisfied, the scheme may show numerical features not related to the underlying continuous equations [16,18]. Through the different sections, we will show how some schemes that satisfy a discrete version of conditions (1)–(3) define appropriate algorithms to solve, in a standard way, general optimization problems associated with the motion of Newtonian particles.

1.2 One-Dimensional Motion of a Newtonian Particle

The equation

$$m\frac{\mathrm{d}^2x}{\mathrm{d}t^2} = F(x) \tag{1.1}$$

describes the motion $x(t)$ of a particle with mass m under a static conservative force $F(x) = -\mathrm{d}U/\mathrm{d}x$, where $U(x)$ is the potential energy. We will suppose that $U(x)$ and $F(x)$ are both smooth functions. As we mentioned above, the motion of such a particle can be determined considering the following properties [13]:

L. Vázquez and S. Jiménez, *Newtonian Nonlinear Dynamics for Complex Linear and Optimization Problems*, Nonlinear Systems and Complexity 4, DOI 10.1007/978-1-4614-5912-5_1, © Springer Science+Business Media New York 2013

1. The conservation of the energy

$$\frac{dE}{dt} = 0, \quad E = \frac{1}{2}m\left(\frac{dx}{dt}\right)^2 + U(x). \tag{1.2}$$

2. The symmetry under time reversal: $t \rightarrow -t$.
3. The properties of the potential $U(x)$ and, specially, the type and stability character of its extrema.

In fact, (1.1) can be expressed as the planar, Hamiltonian, autonomous dynamical system

$$\begin{cases} m\dfrac{dx}{dt} = p, \\ \dfrac{dp}{dt} = F(x), \end{cases} \tag{1.3}$$

whose critical or equilibrium points are $(x, p) = (\bar{x}, 0)$, with $\bar{x}$ such that $F(\bar{x}) = 0$. Since F is minus the derivative of U, $\bar{x}$ corresponds to an extremal point of U. On the other hand, the energy (1.2) becomes:

$$\frac{dE}{dt} = 0, \quad E = \frac{p^2}{2m} + U(x). \tag{1.4}$$

It can be used as a Liapunov function to show the stability of critical points (centres) for which $\bar{x}$ corresponds to a minimum of U. The critical points obtained for maxima of $U(x)$ are unstable (saddle points). If F is sufficiently smooth, the theory for this kind of systems establishes these properties of the motion [24].

In some cases we will solve Eq. (1.1) analytically. In others we will use numerical methods. In order to simulate it we will use methods that satisfy, as much as possible, these conditions. But we will keep to the simpler ones. In this way, we will use, basically, two schemes and their modifications or extensions. Both methods satisfy the symmetry under time reversal, preserve the stability character of the equilibrium points and have a good approximation of the conservation law for the energy.

The simplest one to represent a second order equation, such as (1.1), is the *Størmer–Verlet* [16] scheme, given by

$$m\frac{x_{n+1} - 2x_n + x_{n-1}}{\tau^2} = F(x_n). \tag{1.5}$$

Here τ is the mesh size of the time variable and x_n denotes the computed position at time $t = n\tau$. The symmetry of this expression about x_n corresponds to time reversibility. As for the energy, for the case of a quadratic potential $U(x) = a +$

$bx + cx^2$ and a force $F(x) = -b - 2cx$, there is a discrete conservation law and the following expression is exactly preserved under the discrete dynamics:

$$E_{n+1} = \frac{1}{2}m \left(\frac{x_{n+1} - x_n}{\tau} \right)^2 + a + b \frac{x_{n+1} + x_n}{2} + cx_{n+1}x_n. \qquad (1.6)$$

It can be established multiplying (1.5) by $\dfrac{x_{n+1} - x_{n-1}}{2\tau}$:

$$m\frac{x_{n+1} - x_{n-1}}{2\tau} \frac{x_{n+1} - 2x_n + x_{n-1}}{\tau^2} = \frac{x_{n+1} - x_{n-1}}{2\tau}(-b - 2cx_n)$$

$$\Longleftrightarrow \frac{m}{2\tau} \left(\frac{x_{n+1}^2 - 2x_{n+1}x_n + x_n^2}{\tau^2} - \frac{x_n^2 - 2x_nx_{n-1} + x_{n-1}^2}{\tau^2} \right)$$

$$= -b\frac{x_{n+1} - x_{n-1}}{2\tau} - 2c\frac{x_{n+1}x_n - x_nx_{n-1}}{2\tau}$$

$$\Longleftrightarrow \frac{E_{n+1} - E_n}{\tau} = 0. \qquad (1.7)$$

For any other case, there is no similar law, although the associated evolution map is symplectic, or "canonical", and can be understood as a motion corresponding to a time-dependent Hamiltonian system, "close" to that of the original equation [16]. In order to determine this, we need to use a discrete version of (1.3), with a discrete momentum p_n. For instance:

$$\begin{cases} m\dfrac{x_{n+1} - x_n}{\tau} = p_{n+1}, \\ \dfrac{p_{n+1} - p_n}{\tau} = F(x_n). \end{cases} \qquad (1.8)$$

This system of discrete equations supposes an implicit evolution map from time step n to $n+1$ given by $(x_{n+1}, p_{n+1}) = f(x_n, p_n)$. In this two-variable case, the map is symplectic if its Jacobian has determinant equal to 1. The Jacobian is:

$$Df = \begin{pmatrix} 1 & -\frac{\tau}{m} \\ 0 & 1 \end{pmatrix}^{-1} \begin{pmatrix} 1 & 0 \\ \tau F'(x_n) & 1 \end{pmatrix} = \begin{pmatrix} 1 + \frac{\tau^2}{m}F'(x_n) & \frac{\tau}{m} \\ \tau F'(x_n) & 1 \end{pmatrix}, \qquad (1.9)$$

and $|Df| = 1$. This implies that

$$\frac{p_n^2}{2m} + U(x_n) \qquad (1.10)$$

is "close" to the value of the continuous energy defined in (1.4).

A more sophisticated numerical method is the *Strauss–Vázquez* scheme, expressed by

$$m\frac{x_{n+1} - 2x_n + x_{n-1}}{\tau^2} = -\frac{U(x_{n+1}) - U(x_{n-1})}{x_{n+1} - x_{n-1}}. \tag{1.11}$$

The expression on the right-hand side may (and should) be simplified, for instance, whenever U is a polynomial function. In the case the simplification cannot be carried, the quotient must be replaced by the limit whenever the denominator is of the order of the machine precision, to avoid a small denominator problem.

This method has an exact conservation law for a discrete counterpart of the energy [18]. Multiplying equation (1.11) by $(x_{n+1} - x_{n-1})/2\tau$ and rearranging terms, we get a consistent discretization of the conservation of the energy, in a similar way as we have done for the previous method. Here we have:

$$\frac{E_{n+1} - E_n}{\tau} = 0, \quad E_{n+1} = \frac{1}{2}m\left(\frac{x_{n+1} - x_n}{\tau}\right)^2 + \frac{1}{2}\Big(U(x_{n+1}) + U(x_n)\Big). \tag{1.12}$$

This method is not symplectic: the evolution map implicitly defined by (1.11) does not correspond to a canonical transformation, for any definition of a discrete momentum p_n [17]. Unless the time step τ is allowed to variate, there is not a possibility to have a numerical method that ensures for a general potential both the conservation law of the energy and the symplectic character of the evolution map [16].

From a computational point of view, Størmer–Verlet scheme is explicit while Strauss–Vázquez scheme is implicit and requires solving a simple functional equation for the unknown at each time step. The local truncation error for both methods is $\mathcal{O}(\tau^2)$ and the solution x_{n+1} is computed with an $\mathcal{O}(\tau^4)$ local error.

1.3   One-Dimensional Motion with Linear Dissipation

In a dissipative system the total energy is not conserved but varies in a certain manner depending on the dissipation term. As an illustration of this, let us consider a particle moving under the action of a conservative force and subject to a linear drag force

$$m\frac{d^2x}{dt^2} = F(x) - \alpha\frac{dx}{dt}. \tag{1.13}$$

If the force has to correspond to a dissipation, α must be a positive constant. The variation of the total energy is given by the equation:

$$\frac{dE}{dt} = -\alpha\left(\frac{dx}{dt}\right)^2, \tag{1.14}$$

where the energy E is defined as in (1.2). If $\alpha > 0$, the energy of the particle decreases with time, while it increases for $\alpha < 0$ and, thus, a dissipative system corresponds to the case $\alpha > 0$. If $U(x)$ is bounded from below, the trajectory of the particle will always reach a minimum of the potential U, either global or local, that is an equilibrium point of the system. Seen as a dynamical system, the dissipation transforms the centres of (1.1) into asymptotically stable foci, while the saddle points remain. This is the basis of our method for the estimation of unconstrained extrema of functions, and we will present some examples in Sect. 1.5. The application to constrained extrema, which are related to nonlinear optimization, will be considered in Chap. 7.

A natural extension of the previous numerical methods to the dissipative case is as follows. For the Størmer–Verlet Scheme:

$$m\frac{x_{n+1} - 2x_n + x_{n-1}}{\tau^2} = F(x_n) - \alpha\frac{x_{n+1} - x_{n-1}}{2\tau}, \tag{1.15}$$

and for the Strauss–Vázquez one:

$$m\frac{x_{n+1} - 2x_n + x_{n-1}}{\tau^2} = -\frac{U(x_{n+1}) - U(x_{n-1})}{x_{n+1} - x_{n-1}} - \alpha\frac{x_{n+1} - x_{n-1}}{2\tau}. \tag{1.16}$$

The discrete energies, as defined in Sect. 1.2 by Eqs. (1.6) and (1.12), are now not conserved but satisfy the variation law

$$\frac{E_{n+1} - E_n}{\tau} = -\alpha\left(\frac{x_{n+1} - x_{n-1}}{2\tau}\right)^2, \tag{1.17}$$

which is a consistent discretization of (1.14) and of the dissipative effect. The two variations laws can be established in a very similar way as we did before for the conservative case. In this sense, the algorithms of Størmer–Verlet, for a quadratic potential, and Strauss–Vázquez, in the general case, are suitable to approach the solution of problems with unconstrained extrema. Whenever the problem allows us, we will prefer the Størmer–Verlet method for its simplicity. Nevertheless, in the case of a constant or a linear force (i.e., the cases of the constant gravitational field and the quadratic potential), both methods are equally suitable and, in fact, the Strauss–Vázquez scheme becomes explicit.

1.4 General Motion of a Particle in q Dimensions

Now, the equation of the motion is

$$m\frac{d^2\vec{x}}{dt^2} = \vec{F}(\vec{x}), \tag{1.18}$$

where $\vec{F}$ and $\vec{x}$ are vectors in $\mathbb{R}^q$ and $\vec{F}(\vec{x}) = -\vec{\nabla}U(\vec{x})$, U being the potential energy. We have similar motion properties as in the one-dimensional case. This is the case for the energy conservation

$$\frac{dE}{dt} = 0, \quad E = \frac{1}{2}m\left(\frac{d\vec{x}}{dt}\right)^2 + U(\vec{x}). \tag{1.19}$$

On the other hand, if the motion is under a linear drag force, the equation of motion becomes

$$m\frac{d^2\vec{x}}{dt^2} = \vec{F}(\vec{x}) - \alpha\frac{d\vec{x}}{dt}, \tag{1.20}$$

where α is a constant. The total energy variation is given, quite similar to the one-dimensional case, by:

$$\frac{dE}{dt} = -\alpha\left(\frac{d\vec{x}}{dt}\right)^2. \tag{1.21}$$

We have used the simplified notation, widely used in Mechanics, that represents the 2-norm squared of a vector by its square: $\vec{v}^2 = \|\vec{v}\|^2$.

As before, if $\alpha > 0$, the energy of the particle decreases, while it increases for $\alpha < 0$. For the computation of extrema of a potential $U(\vec{x})$, the considerations made for the one-dimensional case are also valid in this q-dimensional motion. This is a general property since it can be shown that for a Hamiltonian system (without dissipation, thus) given by

$$H = \frac{m}{2}\left(\frac{d\vec{x}}{dt}\right)^2 + U(\vec{x}), \tag{1.22}$$

with $U(\vec{x})$ sufficiently regular, all the critical points are either saddles or centres [13]. Introducing the dissipation factor changes all the centres that correspond to minima of the potential into asymptotically stable foci, while the remaining points are all unstable.

There is a natural extension of the previous numerical methods to the q-dimensional motion to the dissipative case. We have for the Størmer–Verlet scheme:

$$m\frac{\vec{x}_{n+1} - 2\vec{x}_n + \vec{x}_{n-1}}{\tau^2} = \vec{F}(\vec{x}_n) - \alpha\frac{\vec{x}_{n+1} - \vec{x}_{n-1}}{2\tau}, \tag{1.23}$$

which is explicit. For a non-dissipative system ($\alpha = 0$), as in the one-dimensional case, it only shows a discrete conserved energy for quadratic potentials, but the corresponding evolution map, once we express the equations in terms of a discrete momentum vector $\vec{p}_n$, is symplectic.

For the Strauss–Vázquez, we have:

$$m\frac{(x_{n+1})_i - 2(x_n)_i + (x_{n-1})_i}{\tau^2} = -\frac{\Delta_i U(\vec{x}_{n+1}, \vec{x}_{n-1})}{(x_{n+1})_i - (x_{n-1})_i}$$

$$-\alpha\frac{(x_{n+1})_i - (x_{n-1})_i}{2\tau}, \quad i = 1,\dots,q, \tag{1.24}$$

where $(x_n)_i$ is the ith-component of vector $\vec{x}_n$, and $\Delta_i U(\vec{x}_{n+1}, \vec{x}_{n-1})$ is a suitable, symmetric representation of the variation with respect to x_i of the potential $U(\vec{x})$ at time step n. The general expression is rather complicated (see Exercise 1.6 below for $q = 2$, and also [18] for $q = 2$ and $q = 3$) but it simplifies in many cases, for instance, when U is the sum of local potentials (as it is the case for a chain of oscillators with an on-site potential). At each time step a system of q coupled nonlinear equations must be solved, each one associated with the unknown x_i. There is a variation law for a discrete energy, similar to the one-dimensional case of Sects. 1.2 and 1.3.

A simpler numerical scheme that preserves the same energy can be build, in which the equations can be solved sequentially, but the accuracy is one order less in τ (see, again, Exercise 1.6 below, [18] and references therein).

Since we will apply this methods to solving linear and quadratic programming problems, we present now the two following particular cases: a constant gravitational field and a quadratic potential.

1.4.1   Constant Gravitational Field

In this first case, the potential energy is $U = \vec{c}^T \vec{x}$ where $\vec{c} \in \mathbb{R}^q$ is a given vector whose components are, with a minus sign, those of the force of gravity and can be selected as positive by a convenient choice of the reference axis. The equation of motion is

$$m \frac{d^2 \vec{x}}{dt^2} = -\vec{c}. \tag{1.25}$$

The solution of the equation of motion is:

$$\vec{x} = \vec{x}(0) + \frac{d\vec{x}}{dt}(0)t - \frac{1}{2m}\vec{c}t^2. \tag{1.26}$$

Thus, independently of the initial conditions, for large t, the components of $\vec{x}$ tend quadratically in time towards $-\infty$, since the potential is unbounded from below. Essentially, this is what happens to the metallic ball in the pinball machine, and it will be the basis for our method to compute the solution of the optimization of a linear objective function with linear or nonlinear constraints. This will be discussed in Chaps. 6 and 7.

1.4.2   Quadratic Potential

In this second case, let be the potential

$$U(\vec{x}) = a - \vec{c}^T \vec{x} + \frac{1}{2}\vec{x}^T A \vec{x}, \tag{1.27}$$

where $a \in \mathbb{R}$, $\vec{c} \in \mathbb{R}^q$ and A is a $q \times q$ symmetric and positive definite real matrix. Finding the minimum of this potential reduces to solve the linear problem $A\vec{x} = \vec{c}$. This can be done either by algebraic classical methods or by a mechanical approach, using the motion of a particle described above: it will be developed in detail in Chaps. 2 and 3. The case with constraints will be studied in the context of the quadratic programming in Chap. 7.

1.5 Unconstrained Extrema

Let us consider the problem of finding the extrema of a scalar real function $f(\vec{x})$ of q real variables $\vec{x} = (x_1, \ldots, x_q)$. More precisely, let us consider only the minima, since the maxima correspond to the minima of the function $-f(\vec{x})$. If we identify the function $f(\vec{x})$ with the potential $U(\vec{x})$ that a Newtonian particle "sees", the local minima of the function correspond to the stable equilibrium points of the motion. Thus, we can apply the algorithms of Sects. 1.3 and 1.4 to approximate the minima. We illustrate this in what follows for several cases.

Example 1.5.1 (One Variable).
   Let us consider finding the minima of the two functions

$$f(x) = \frac{x^2}{1+x^2}, \quad g(x) = \frac{x^4 - x^3 - x^2}{1 + x^2 + x^4}. \tag{1.28}$$

Function $f(x)$ has a local minimum at $x = 0$ while function $g(x)$ has a local minimum located at $x_{\min} \approx -0.4005065791469695$ and an absolute minimum located at $x_{\text{Min}} \approx 0.8310632618394241$, as can be seen from Fig. 1.1.
If we apply the Størmer–Verlet and Strauss–Vázquez methods using the function $f(x)$ as potential, we obtain the minimum $x = 0$, $\dot{x} = 0$, in the phase space. The errors of both methods are presented in Fig. 1.2 for the values: $m = 1$, $\tau = 0.1$ and $\alpha = 2.825$.
   In the case of function $g(x)$, we have represented the errors of both methods in Fig. 1.3. The two minima are obtained using different initial values.

Example 1.5.2 (Quadratic functions of two variables).
   Let us consider a quadratic potential of the form

$$U(x,y) = x^2 + y^2 - 3x - xy + 3, \tag{1.29}$$

whose minimum is located at $(2,1)$. We apply the dissipative Størmer–Verlet scheme (1.23) to obtain that minimum. We will see in Chap. 2 how to choose parameters τ, m and α in such a way as to optimize the method. In this example we have: $m = 1/4$, $\alpha = \sqrt{3}/2$ and $\tau \in (0,1)$. In Fig. 1.4 we represent for these values and $\tau = 1/2$, the error versus the number of iterations. After 28 iterations the value is exact up to machine precision.

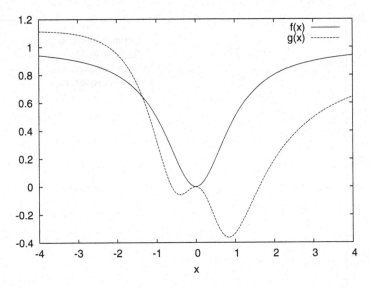

Fig. 1.1   Example 1.5.1. Minima of functions of one variable

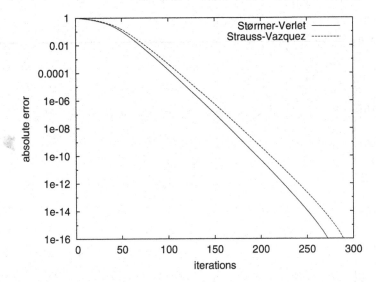

Fig. 1.2   Example 1.5.1. Errors of the methods. Function $f(x)$

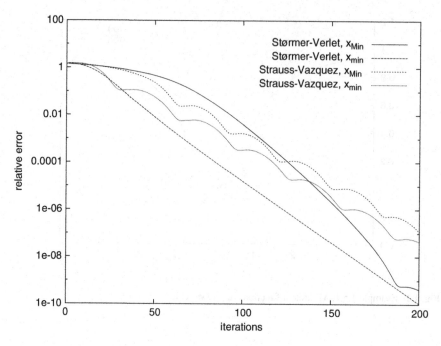

Fig. 1.3 Example 1.5.1. Errors of the methods. Function $g(x)$

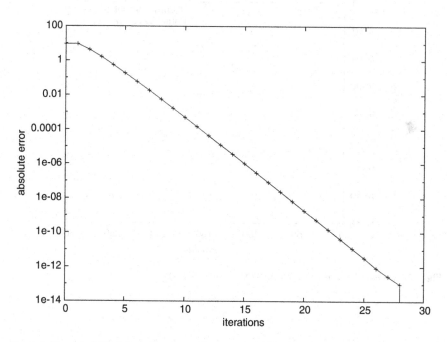

Fig. 1.4 Example 1.5.2. Minimum of a quadratic potential in two variables

1.6 Exercises

1.1 Check the variation law of the discrete energy for the one-dimensional dissipative Størmer–Verlet and Strauss–Vázquez schemes, given by (1.17) in both cases.

1.2 Using your favourite programming language, implement the dissipative Størmer–Verlet numerical method, in many dimensions. Apply it to locate the minima of the two variables function:

$$f(x,y) = -(x^2 + y^2) + (x^2 + y^2 - 4x)^2. \tag{1.30}$$

1.3 The expression of the Strauss–Vázquez scheme, conservative case, in two dimensions is given in the symmetrized version by:

$$\frac{x_{n+1} - 2x_n + x_{n-1}}{\tau^2} + \frac{U(x_{n+1}, y_{n-1}) - U(x_{n-1}, y_{n-1})}{2(x_{n+1} - x_{n-1})}$$

$$+ \frac{U(x_{n+1}, y_{n+1}) - U(x_{n-1}, y_{n+1})}{2(x_{n+1} - x_{n-1})} = 0, \tag{1.31}$$

$$\frac{y_{n+1} - 2y_n + y_{n-1}}{\tau^2} + \frac{U(x_{n-1}, y_{n+1}) - U(x_{n-1}, y_{n-1})}{2(y_{n+1} - y_{n-1})}$$

$$+ \frac{U(x_{n+1}, y_{n+1}) - U(x_{n+1}, y_{n-1})}{2(y_{n+1} - y_{n-1})} = 0. \tag{1.32}$$

The unsymmetrized expression is:

$$\frac{x_{n+1} - 2x_n + x_{n-1}}{\tau^2} + \frac{U(x_{n+1}, y_{n-1}) - U(x_{n-1}, y_{n-1})}{x_{n+1} - x_{n-1}} = 0, \tag{1.33}$$

$$\frac{y_{n+1} - 2y_n + y_{n-1}}{\tau^2} + \frac{U(x_{n+1}, y_{n+1}) - U(x_{n+1}, y_{n-1})}{y_{n+1} - y_{n-1}} = 0. \tag{1.34}$$

Although less accurate, this second expression is easier to implement since the first equation does not depend on the variable y_{n+1}, and the equations can be solved sequentially instead of simultaneously.

Check that in both cases the discrete energy

$$E_n = \frac{1}{2}\left[\left(\frac{x_{n+1} - x_n}{\tau}\right)^2 + \left(\frac{y_{n+1} - y_n}{\tau}\right)^2\right]$$

$$+ \frac{U(x_{n+1}, y_{n+1}) + U(x_n, y_n)}{2}, \tag{1.35}$$

is exactly conserved.

1.4 (a) Find the variation law for the energy, defined by (1.35), for the two-dimensional dissipative Strauss–Vázquez scheme (either symmetrized or not).

  (b) Compare it with the continuous one, given by (1.21).

1.5 (a) Implement the two-dimensional dissipative Strauss–Vázquez scheme (either symmetrized or not) in your favourite programming language.

  (b) Use it to locate the minima of function (1.30). Compare with Exercise 1.6.

  (c) Check, numerically, in your code, the variation law for the energy (see Exercises 1.6 and 1.6).

1.6 Let be the two variables function

$$f(x,y) = \left(x^2 - y^2 + \frac{x}{2} + \frac{y}{4}\right) e^{-x^2 - y^2/2}.$$

(a) Find numerically the minima, using both dissipative methods: the Størmer–Verlet and the Strauss–Vázquez.

(b) Find also the maxima, as minima of function $-f$.

1.7 Repeat the previous exercise, now for the function

$$f(x,y) = -\ln(1 + x^2 + 2y^2 - x)\, e^{-x^2 - y^2/2}.$$

1.8 Repeat again, now for

$$f(x,y) = (-12xy + 3x + y)/(1 + x^2 + y^2)^2.$$

1.9 Find numerically the minima of function:

$$f(x,y) = \frac{x^2}{4}(x-1)^2 - \frac{x^2}{2} + \left(\frac{y^3}{2} - y + 2\right)^2,$$

using both dissipative methods. Compare with the analytical solution.

1.10 Find numerically the minima of function:

$$f(x,y) = \frac{x^2}{5}(x-1)(x-3)\sqrt{x^2 + y^2} + \frac{y^4}{4} - \frac{(y-1)^2}{2}.$$

1.11 A discrete evolution map given by $(\vec{x}_{n+1}, \vec{p}_{n+1}) = h(\vec{x}_n, \vec{p}_n)$, with $\vec{x}_n, \vec{p}_n \in \mathbb{R}^q$, is symplectic, or canonical if its Jacobian Dh satisfies the relation: $(Dh)^T J Dh = J$, where J is the matrix given in block-form by:

$$J = \left(\begin{array}{c|c} O & I \\ \hline -I & O \end{array}\right)$$

and I and O are, respectively, the identity and the null matrices of order q.

(a) Consider the Størmer–Verlet method with $q = 2$, in the conservative case, and check that the corresponding evolution map written as a system, in a similar way to what was done for (1.8), is indeed symplectic.

(b) Following what you have done in part (a), show now that the general method, with q arbitrary, is symplectic.

1.12 The Strauss–Vázquez method can be understood as preserving at discrete level the chain rule. For instance, in the one-dimensional case we have

$$\frac{dU(x)}{dt} = \frac{\partial U(x)}{\partial x}\frac{dx}{dt}$$

expressed as:

$$\frac{1}{\tau}\left(\frac{U(x_{n+1})+U(x_n)}{2} - \frac{U(x_n)+U(x_{n-1})}{2}\right)$$

$$= \frac{U(x_{n+1})+U(x_{n-1})}{x_{n+1}-x_{n-1}}\frac{x_{n+1}-x_{n-1}}{2\tau}.$$

Keeping this in mind, and using (1.33) and (1.34) as references, build the numerical scheme for the three-dimensional case. You may just consider a simpler, unsymmetrized expression. You may check the expression you obtain with that in [18]. As an additional reference, the expression of the conserved energy (non-dissipative case) is:

$$E_n = \frac{1}{2}\left[\left(\frac{x_{n+1}-x_n}{\tau}\right)^2 + \left(\frac{y_{n+1}-y_n}{\tau}\right)^2 + \left(\frac{z_{n+1}-z_n}{\tau}\right)^2\right]$$

$$+ \frac{U(x_{n+1},y_{n+1},z_{n+1})+U(x_n,y_n,z_n)}{2}. \tag{1.36}$$

1.13 (a) Implement the dissipative Strauss–Vázquez method in three dimensions, following the ideas of the previous example.

(b) Use it, together with the dissipative Størmer–Verlet scheme, to locate the minimum of the function:

$$f(x,y,z) = 2x^2 + 2y^2 + z^2 + 2xy + 2xz + 9 - 8x - 6y - 4z.$$

Compare with the analytical solution.

1.14 Find, numerically, the minima of the function:

$$f(x,y,z) = \frac{x^2 + 2xyz - xz^2}{(1+x^2+2y^2+z^2)^2}$$

Chapter 2
Solution of Systems of Linear Equations

2.1 Introduction

Solving systems of linear equations (or *linear systems* or, also, *simultaneous equations*) is a common situation in many scientific and technological problems. Many methods, either analytical or numerical, have been developed to solve them. A general method most used in Linear Algebra is the Gaussian Elimination, or variations of this. Sometimes they are referred to as "direct methods". Basically, it is an algorithm that transforms the system into an equivalent one but with a triangular matrix, thus allowing a simpler resolution. In many cases, though, whenever the matrix of the system has a specific structure or is sparse and the like, other methods can be more effective. They are "iterative methods", not based on a finite algorithm but on an iterative process. The two simpler methods of this kind are the Jacobi and the Gauss–Seidel methods. There is also an iterative method which transforms solving the system into minimizing a scalar function: the "Method of the Steepest Descent".

In our case, we present here a simple iterative method based on the motion of a damped harmonic oscillator in a linear force field plus a constant force field, such as the gravitational field on the surface of the Earth [13, 31]. This mechanical system evolves, under the action of time, towards the solution of the linear equations. We will apply very similar ideas to compute eigenvectors of matrices and to solve linear and nonlinear programming problems in further chapters. In the next one we will compare the performance of the mechanical method with that of the iterative methods mentioned above.

To illustrate the basic ideas, let us consider the one-dimensional case. We want to solve the (trivial, in one dimension) linear equation

$$ax = b. \tag{2.1}$$

To do this, we will consider a mechanical system whose solution tends to the solution of this equation. The equation of motion for a particle under a linear force

L. Vázquez and S. Jiménez, *Newtonian Nonlinear Dynamics for Complex Linear and Optimization Problems*, Nonlinear Systems and Complexity 4, DOI 10.1007/978-1-4614-5912-5_2, © Springer Science+Business Media New York 2013

and a constant external acceleration is

$$m\frac{d^2x}{dt^2} + \alpha\frac{dx}{dt} + ax = b,\qquad(2.2)$$

where $x = x(t)$ is the one-dimensional displacement of the mass m under a dissipation ($\alpha > 0$), a harmonic force with $a > 0$, and a constant acceleration (b, due to a gravitational field, for instance). From the theory of linear differential equations, we know that the general solution to (2.2) can be expressed as the sum of two contributions, the *general solution of the homogeneous part*, x_h, plus a *particular solution*, x_p:

$$x(t) = x_h(t) + x_p(t).\qquad(2.3)$$

Moreover, since

$$\lim_{t\to\infty} x_h(t) = 0,\qquad \lim_{t\to\infty} x_p(t) = \frac{b}{a},\qquad(2.4)$$

the solution can be expressed as a sum of a time-dependent term, $x_0(t)$, plus a constant one:

$$x(t) = x_0(t) + \frac{b}{a},\qquad(2.5)$$

with $x_0(t)$ decaying exponentially to zero as t goes to infinity (if both α and a are positive, as we suppose). The specific rate of that exponential decay depends on the coefficients in Eq. (2.2), but in every case, and independently of the initial conditions, the solution tends to the *asymptotically stable* constant solution given by $x(t) = b/a$.

From a mechanical point of view, the system has a total energy given by

$$E = \frac{m}{2}\left(\frac{dx}{dt}\right)^2 + \frac{1}{2}ax^2 - bx,\qquad(2.6)$$

with variation law

$$\frac{dE}{dt} = -\alpha\left(\frac{dx}{dt}\right)^2.\qquad(2.7)$$

This implies, α being positive, that E will decrease under the evolution. Since E is bounded from below by the minimum value $-b^2/2a$, the system will approach asymptotically a state given by

$$\frac{dx}{dt} = 0,\qquad x = \frac{b}{a}.\qquad(2.8)$$

We see that both approaches, this Mechanical one and solving directly Eq. (2.1), give the same result. In fact, the equivalence of both methods is part of Liapunov stability theory.

Remarks. 1. In the *overdamped* regime ($\alpha > 4am$), when α is large enough as compared to a and m, the dominant effect is the dissipative one and the equation of motion can be approximated by the first order equation:

$$\alpha \frac{dx}{dt} + ax = b, \tag{2.9}$$

whose solutions have the same asymptotic behaviour as those of (2.2), and tend to the constant solution $x = b/a$, independently of the initial conditions. The presence of coefficient α can be avoided rescaling time.

2. If we allow a to be zero in (2.2), the behaviour drastically changes and, in general, *secular* terms appear that are neither constant nor vanishing. If both, a and b, are zero, the system evolves again into a constant solution that depends now on the initial conditions. The same happens with (2.9). In the general case of many dimensions, this corresponds to undetermined systems of linear algebraic equations.

2.2 General Case

The mechanical ideas considered previously are the bases for the two families of iterative methods that we present to solve a system of linear algebraic equations. The methods are constructed choosing a finite difference method to solve either one of the associated linear differential equations, (2.2) or (2.9). In the next two sections, we will present the general case, extending the previous basic, one-dimensional, mechanical considerations to the case of a system of equations.

In this approach, we translate the problem of solving a system of linear equations into solving the equation of motion of one particle that tends asymptotically to a position which is identified with the solution of that linear system. Because of this asymptotic behaviour, we can expect to have general iterative methods which need many iterations to approach the solution, but the convergence is satisfied whenever the discretization of the differential equation of motion satisfies conditions related to the conservation and variation of the energy of the system.

Let be the linear system

$$A\vec{x} = \vec{b}, \tag{2.10}$$

where we assume that A is an $q \times q$ non-singular matrix (i.e., the system has a unique solution). We may associate to it Newton's equation for a linear dissipative ($\alpha > 0$) mechanical system:

$$\frac{d^2\vec{x}}{dt^2} + \alpha \frac{d\vec{x}}{dt} + A\vec{x} = \vec{b} \tag{2.11}$$

and the overdamped asymptotic equation

$$\frac{d\vec{x}}{dt} + A\vec{x} = \vec{b}. \tag{2.12}$$

For both equations, if A has a real, positive definite spectrum we have

$$\lim_{t \to \infty} \vec{x}(t) = A^{-1}\vec{b}, \qquad (2.13)$$

which is the solution of the linear system (2.10). Since A is not necessarily a positive definite matrix, we may consider an equivalent problem, given by

$$\frac{d^2\vec{x}}{dt^2} + \alpha \frac{d\vec{x}}{dt} + M\vec{x} = \vec{v} \qquad (2.14)$$

or

$$\frac{d\vec{x}}{dt} + M\vec{x} = \vec{v} \qquad (2.15)$$

in which:

$$M\vec{x} = \vec{v} \iff A\vec{x} = \vec{b}, \qquad (2.16)$$

and such that M is positive definite. Depending on the properties of A, we may choose different possibilities for M and $\vec{v}$ (see [32]):

(1) When the spectrum of A is real and positive: $\ddot{\vec{x}} + \alpha \dot{\vec{x}} + A\vec{x} = \vec{b}$.
(2) When the spectrum of A is real and negative: $\ddot{\vec{x}} + \alpha \dot{\vec{x}} - A\vec{x} = -\vec{b}$.
(3) When the spectrum of A is real: $\ddot{\vec{x}} + \alpha \dot{\vec{x}} + AA\vec{x} = A\vec{b}$.
(4) When A has complex eigenvalues: $\ddot{\vec{x}} + \alpha \dot{\vec{x}} + A^T A\vec{x} = A^T \vec{b}$.

We have make use of the "dotted" notation to represent the time derivatives in a more compact way.

In order to avoid problems with the spectrum of A, which is in principle not known beforehand, we will consider in what follows

$$M = A^T A, \quad \vec{v} = A^T \vec{b}. \qquad (2.17)$$

Although this may not be a good idea if A is ill-conditioned (see Exercise 2.5, below, and also the discussion of formula (2.7.40) in page 85 of [25]), we, thus, ensure that M is symmetric and positive definite by construction and has, therefore, a real, positive definite spectrum, and that the constant solution $\vec{x}(t) = M^{-1}\vec{v} = A^{-1}\vec{b}$ is asymptotically stable.

As before, we can use the existence of a decreasing energy for (2.14):

$$E(\vec{x}) = \frac{1}{2}\frac{d\vec{x}^T}{dt}\frac{d\vec{x}}{dt} + \frac{1}{2}\vec{x}^T M\vec{x} - \vec{x}^T \vec{v},$$

$$= \frac{1}{2}\left\|\frac{d\vec{x}}{dt}\right\|^2 + \frac{1}{2}\vec{x}^T M\vec{x} - \vec{x}^T \vec{v}, \qquad (2.18)$$

where $\|\cdot\|$ is the Euclidean vector norm, or 2-norm. The variation law is

$$\frac{dE(\vec{x})}{dt} = -\alpha\left\|\frac{d\vec{x}}{dt}\right\|^2. \qquad (2.19)$$

Once the evolution equation is chosen, we integrate it with a numerical scheme in order to approach the asymptotically stable solution. Since we have exchanged our problem for an equivalent one, we must keep in mind that our method should never be more time-consuming than tackling the original problem: we look thus for an explicit, finite-difference method to solve the equations. We have chosen the forward Euler method. We will call it the *damped* method when applied to (2.14) and the *overdamped* one when applied to (2.15).

2.3 Damped Method

A suitable, finite difference scheme to solve (2.14) is the following [22, 29]:

$$\frac{\vec{x}_{n+1} - 2\vec{x}_n + \vec{x}_{n-1}}{\tau^2} + \alpha \frac{\vec{x}_{n+1} - \vec{x}_{n-1}}{2\tau} + M\vec{x}_n = \vec{v}, \qquad (2.20)$$

where τ is the mesh-size of the time variable and $\vec{x}_n$ denotes the position at time $t = n\tau$. We have chosen it since it is the simplest finite difference method to solve (2.14). The scheme is a two-step recursion, as it should since it represents a second order equation, and here we need two initial values to start the computations. Since we are looking for an asymptotically stable solution, we may choose, for instance, $\vec{x}_1 = \vec{x}_0$ and $\vec{x}_0 \neq \vec{0}$ arbitrary. To actually perform the numerical computations, we may express the scheme as

$$\vec{x}_{n+1} = \frac{1}{1 + \frac{\tau\alpha}{2}} \left[(2I - \tau^2 M) \vec{x}_n - \left(1 - \frac{\tau\alpha}{2}\right) \vec{x}_{n-1} + \tau^2 \vec{v} \right], \qquad (2.21)$$

where I denotes the identity matrix of the appropriate order. If we multiply (2.20) on the left by $(\vec{x}_{n+1} - \vec{x}_{n-1})^{\mathrm{T}}/2\tau$, and rearrange terms, we obtain

$$\frac{E_{n+1} - E_n}{\tau} = -\alpha \left(\frac{\vec{x}_{n+1} - \vec{x}_{n-1}}{2\tau} \right)^2, \qquad (2.22)$$

where

$$E_{n+1} \equiv \frac{1}{2} \left(\frac{\vec{x}_{n+1} - \vec{x}_n}{\tau} \right)^2 + \frac{1}{2} \vec{x}_n^{\mathrm{T}} M \vec{x}_{n+1} - \frac{1}{2} \left(\vec{x}_{n+1}^{\mathrm{T}} \vec{v} + \vec{x}_n^{\mathrm{T}} \vec{v} \right) \qquad (2.23)$$

is the discrete counterpart of the energy (2.18).

As we can see, the variation of the discrete energy associated with the difference equation (2.20) is similar to that in (2.19), and its decreasing character depends only on the sign of α and not on the solution. This property guarantees the convergence of the numerical solution to that of the system (2.10), provided the value of τ is small enough, unless it has no solution. If that is the case, Eq. (2.14) has a linear component that grows linearly in time and the solution of (2.20) does not converge to a constant.

Equation (2.20) has two arbitrary parameters, α and τ. We will see in what follows how to choose them in order to optimize the computations.

Although a single equation such as (2.21) is more accurate, computation-wise, for the sake of the analysis we translate (2.20) into a system of two equations. Keeping in mind the Mechanical analogy we define:

$$\vec{p}_n = \frac{\vec{x}_{n+1} - \vec{x}_n}{\tau},$$

(2.24)

which is a consistent discrete representation of the momentum (since $m = 1$). With this and (2.20) the scheme becomes

$$
\begin{cases}
\left(\frac{\alpha}{2}I + \tau M\right)\vec{x}_{n+1} + \left(1 + \frac{\tau\alpha}{2}\right)\vec{p}_{n+1} = \frac{\alpha}{2}\vec{x}_n + \vec{p}_n + \tau\vec{v}, \\
\vec{x}_{n+1} = \vec{x}_n + \tau\vec{p}_n,
\end{cases}
$$

(2.25)

where I is the $q \times q$ identity matrix. Let us write this in matrix and vector form as:

$$N_+ \vec{Y}_{n+1} = N_- \vec{Y}_n + \vec{W},$$

(2.26)

with the block matrices and stacked vectors:

$$N_+ = \left(\begin{array}{c|c} \frac{\alpha}{2}I + \tau M & \left(1 + \frac{\tau\alpha}{2}\right)I \\ \hline I & O \end{array}\right), \quad N_- = \left(\begin{array}{c|c} \frac{\alpha}{2}I & I \\ \hline I & \tau I \end{array}\right),$$

(2.27)

$$\vec{Y}_{n+1} = \left(\begin{array}{c} \vec{x}_{n+1} \\ \vec{p}_{n+1} \end{array}\right), \quad \vec{Y}_n = \left(\begin{array}{c} \vec{x}_n \\ \vec{p}_n \end{array}\right), \quad \vec{W} = \left(\begin{array}{c} \tau\vec{v} \\ \vec{0} \end{array}\right),$$

(2.28)

where O is the $q \times q$ null matrix.

It can be easily seen that matrix N_+ is invertible unless $\tau\alpha = -2$, which cannot occur since α and τ are both positive. Thus we have an iterative process that can be written formally as

$$\vec{Y}_{n+1} = (N_+)^{-1}N_-\vec{Y}_n + (N_+)^{-1}\vec{W}.$$

(2.29)

A sufficient condition to ensure the convergence of this process for any initial values is to have all eigenvalues of the iteration matrix

$$N \equiv (N_+)^{-1}N_-$$

(2.30)

of modulus strictly less than 1. Let us compute those eigenvalues. From the characteristic equation, we have (in block form)

$$\lambda \text{ is eigenvalue of } N \iff \left|\begin{array}{c|c} (1-\lambda)\frac{\alpha}{2}I - \lambda\tau M & \left[1 - \lambda\left(1 + \frac{\tau\alpha}{2}\right)\right]I \\ \hline (1-\lambda)I & \tau I \end{array}\right| = 0$$

(and dealing with columns to get an upper triangular block matrix:)

$$\Longleftrightarrow \quad \left| \begin{array}{c|c} -\lambda\tau M + (1-\lambda)\left[\frac{\alpha}{2} - \frac{1}{\tau} + \frac{\lambda}{\tau}\left(1 + \frac{\alpha\tau}{2}\right)\right]I & \left[1 - \lambda\left(1 + \frac{\tau\alpha}{2}\right)\right]I \\ \hline O & \tau I \end{array} \right| = 0$$

$$\Longleftrightarrow \quad \left| M - \frac{1-\lambda}{\lambda\tau}\left[\frac{\alpha}{2} - \frac{1}{\tau} + \frac{\lambda}{\tau}\left(1 + \frac{\alpha\tau}{2}\right)\right]I \right| = 0$$

$$\Longleftrightarrow \quad \frac{1-\lambda}{\lambda\tau}\left[\frac{\alpha}{2} - \frac{1}{\tau} + \frac{\lambda}{\tau}\left(1 + \frac{\alpha\tau}{2}\right)\right] \quad \text{is an eigenvalue of } M,$$

$$\Longleftrightarrow \quad \left(1 + \frac{\alpha\tau}{2}\right)\lambda^2 + (\mu\tau^2 - 2)\lambda + \left(1 - \frac{\alpha\tau}{2}\right) = 0, \tag{2.31}$$

where μ is any eigenvalue of M. Thus, for every eigenvalue μ of M, we get two eigenvalues of N:

$$\lambda_\pm(\mu, \tau, \alpha) = \frac{2 - \mu\tau^2 \pm \tau\sqrt{\mu^2\tau^2 - 4\mu + \alpha^2}}{2 + \alpha\tau}. \tag{2.32}$$

If we want the fastest convergence rate, we should look for values of α and τ such that $|\lambda|$ be as small as possible and, in any case, strictly less than 1: in this way our iterative process will be convergent.

A fundamental property of the second order equation in λ (2.31) is that for any eigenvalue μ of M, we have

$$\lambda_+(\mu, \tau, \alpha)\, \lambda_-(\mu, \tau, \alpha) = \frac{2 - \tau\alpha}{2 + \tau\alpha}, \tag{2.33}$$

and, thus, independent of the value of μ. Since the time step τ is positive, this quantity is less than 1: if we can manage to have $\lambda_\pm$ complex, non-real, it means that both would have a modulus strictly less than 1, since in that case they will be complex conjugate and $\lambda_+\lambda_- = |\lambda_\pm|^2$. Thus the iteration would be convergent. Moreover, we may look for optimal values of τ and α in the following way: let us consider some specific eigenvalue μ. We want:

$$\lambda_+ = \bar{\lambda}_- \text{ (complex, non-real)} \Longleftrightarrow \mu^2\tau^2 - 4\mu + \alpha^2 \leq 0$$

$$\Longleftrightarrow \mu \in [\mu_-, \mu_+], \tag{2.34}$$

where:

$$\mu_- = \frac{2 - \sqrt{4 - \tau^2\alpha^2}}{\tau^2}, \quad \mu_+ = \frac{2 + \sqrt{4 - \tau^2\alpha^2}}{\tau^2}. \tag{2.35}$$

This can be inverted to give:

$$\tau = \frac{2}{\sqrt{\mu_+ + \mu_-}}, \quad \alpha = 2\sqrt{\frac{\mu_+ \mu_-}{\mu_+ + \mu_-}}. \tag{2.36}$$

If we want this to hold for every eigenvalue μ, we may choose μ_- as the smallest eigenvalue of M, and μ_+ as the greatest: these are real positive values since we have chosen $M = A^{\mathrm{T}}A$ and, thus, symmetric and positive definite. In this way we ensure convergence of the method independently of the initial conditions.

In fact, the eigenvalues μ of M are related to the singular values σ of the original matrix A:

$$\mu = \sigma^2. \tag{2.37}$$

We may thus define σ_+ and σ_-, accordingly. Once these values are known, or equivalently μ_+ and μ_-, we compute via (2.36) τ and α and get a rough a priori estimate of the rate of convergence. From (2.32) we have

$$|\lambda| = \sqrt{\frac{2 - \tau\alpha}{2 + \tau\alpha}} = \frac{\sigma_+ - \sigma_-}{\sigma_+ + \sigma_-} \tag{2.38}$$

and we may estimate the error at iteration step n by $|\lambda|^n$. The key point in choosing the most favourable values for parameters τ and α seems, thus, to compute, or at least estimate, μ_+ and μ_-. We present in Chap. 4, and implement in Chap. 5, an iterative method that can be used to achieve this in a reasonable way. Otherwise, tentative values of τ and α may be used. A problem arises whenever μ_- is much smaller than μ_+, since $|\lambda|$ will be very close to unity and the convergence very slow.

Although a priori information about μ_+ and μ_- is not easy to obtain, at least there is a bound that can be established in a simple way: since M is a positive definite matrix, all its eigenvalues are real and strictly positive and, thus, its trace, which corresponds to the sum of all the eigenvalues, is an upper bound to $\mu_+ + \mu_-$. This provides us with the following lower bound for the optimal value of τ:

$$\tau \geq \frac{2}{\sqrt{\mathrm{tr}(M)}}, \tag{2.39}$$

In the case of two dimensions, we have in fact that

$$\mu_+ + \mu_- = \mathrm{tr}(M), \quad \mu_+ \mu_- = \det(M), \tag{2.40}$$

and we can determine the optimal values of both parameters without computing the eigenvalues:

$$\tau = \frac{2}{\mathrm{tr}(M)}, \quad \alpha = 2\sqrt{\frac{\det(M)}{\mathrm{tr}(M)}}. \tag{2.41}$$

We will use this in Example 3.1.1, in next chapter.

2.4 Overdamped Method

To simulate Eq. (2.15) we use the numerical scheme

$$\frac{\vec{x}_{n+1} - \vec{x}_n}{\tau} + M\vec{x}_n = \vec{v}, \tag{2.42}$$

or equivalently

$$\vec{x}_{n+1} = (I - \tau M)\vec{x}_n + \tau\vec{v}. \tag{2.43}$$

This corresponds to the forward Euler Method applied to the overdamped equation (2.15). If we take $\alpha = 2/\tau$ in (2.21), we obtain

$$\vec{x}_{n+1} = \left(I - \frac{\tau^2}{2}M\right)\vec{x}_n + \frac{\tau^2}{2}\vec{v}, \tag{2.44}$$

and we see that this overdamped method is just a particular case of the damped one with an effective time step $\tau' = \tau^2/2$. Since the value of α is fixed with respect to the mesh-size τ, the optimal values obtained in the previous section cannot be applied here in general, so we expect this method to be less efficient than the damped one. Nevertheless, we will perform its analysis: here we do not need an extra variable and the method is convergent simply if the matrix $I - \tau M$ has all its eigenvalues of modulus less than one.

$$\lambda \text{ is eigenvalue of } I - \tau M \iff |I - \tau M - \lambda I| = 0 \tag{2.45}$$

$$\iff \left|M - \frac{1-\lambda}{\tau}I\right| = 0 \iff \frac{1-\lambda}{\tau} \text{ is eigenvalue of } M. \tag{2.46}$$

Again, letting μ be any eigenvalue of M, we have

$$\mu = \frac{1-\lambda}{\tau} \iff \lambda = 1 - \tau\mu. \tag{2.47}$$

Now all λ's are real and the condition for all of them to be of modulus less than one corresponds to

$$0 < \tau < \frac{2}{\mu_+}, \tag{2.48}$$

where μ_+ is the largest eigenvalue of M. This condition by itself is not sufficient to determine the optimal value of τ. In fact, what would be required is to minimize all the possible values of λ. This implies, in turn, knowing, beforehand, all the eigenvalues of M, which is not realistic in the general case. On the other hand, we may estimate the error decay to go as $|\lambda|^n$, n being the number of iterations, which implies that we should try to minimize this value as much as possible. We, thus, may consider as a reference the two following values:

$$\tau_+ = \frac{1}{\mu_+}, \quad \tau_- = \frac{1}{\mu_-}. \tag{2.49}$$

We expect the optimal value of τ, that is, the one that minimizes the number of iterations to achieve a given precision, to lie between these two.

2.5 Singular Matrix

Let be the linear dissipative ($\alpha > 0$) mechanical system:

$$\frac{d^2\vec{x}}{dt^2} + \alpha \frac{d\vec{x}}{dt} + A^T A \vec{x} = \vec{0}. \tag{2.50}$$

If the matrix A is singular, the equation $A\vec{x} = \vec{0}$ has an infinite number of solutions, otherwise only the trivial solution $\vec{x} = \vec{0}$ exists. As a consequence, when numerically computing the solution of the equation above with arbitrary initial conditions, we have two possible behaviours:

- Matrix A is *non-singular*: in this case all numerical solutions converge to $\vec{0}$ as time goes to infinity.
- Matrix A is *singular*: in this case some numerical solution converges to a vector different from $\vec{0}$ as time goes to infinity.

It is clear that the fixed points of the dynamical system correspond to vectors belonging to the kernel of A. The problem is to determine whether some are non-null and use this as a criterion to determine whether a matrix is singular or not.

We will see in the implementations in next chapter how to proceed in a practical way but here we illustrate the two behaviours through the following example.

Example 2.5.1. Let be

$$A = \begin{pmatrix} -1 & -2 & -8 & -12 & 4 \\ -3 & 4 & 8 & 15 & -7 \\ -1 & 3 & 11 & 17 & -7 \\ 1 & -1 & -4 & -6 & 3 \\ 1 & 3 & 12 & 19 & -6 \end{pmatrix}.$$

This is a non-singular matrix, and the iterative process tends to $\vec{0}$, independently of the initial vector chosen.

On the other hand,

$$A = \begin{pmatrix} 2 & -2 & -8 & -9 & 7 \\ -6 & 4 & 8 & 12 & -10 \\ -5 & 3 & 11 & 13 & -11 \\ 2 & -1 & -4 & -5 & 4 \\ -4 & 3 & 12 & 14 & -11 \end{pmatrix}$$

is singular. Starting, for instance, with initial vector $\vec{x}_0 = (1,0,0,0,1)^{\mathrm{T}}$, we obtain numerically convergence towards vector:

$$(-0.243500, 0.243500, 0.324667, -0.0811667, 0.405833)^{\mathrm{T}}.$$

Our result is valid for a general matrix. Thus, we may determine the singular character of a matrix computing numerically some solutions of an associated linear dissipative mechanical system.

A different possibility is to check whether matrix A has a zero eigenvalue. We will address this case in Chap. 5.

2.6 Exercises

2.1 Show that no positive values μ_+ and μ_- exist such that the overdamped method is just the damped one with optimal values for the parameters α and τ.

2.2 (a) Compute M, $\mu_{\pm}$, τ, α and determine the value of $|\lambda|$ for matrix

$$A = \begin{pmatrix} 1 & -2 \\ 1 & -1 \end{pmatrix}.$$

(b) Compute the estimated number of iterations needed to obtain an absolute error $\|A\vec{x}_n - \vec{b}\|$ less that 10^{-12}.

2.3 Matrix A, defined as

$$A = \begin{pmatrix} 1 & 1 \\ 1 & -1 \end{pmatrix},$$

is well conditioned, since the rows correspond to an orthogonal set of vectors.

(a) Compute M, $\mu_{\pm}$, τ, α and determine the value of $|\lambda|$.

(b) Compute the estimated number of iterations needed to obtain an absolute error $\|A\vec{x}_n - \vec{b}\|$ less that 10^{-12}. Is there a significant difference with the result of the previous exercise?

2.4 Repeat the previous exercise using now the matrix

$$A = \begin{pmatrix} 1 & 0 & 1 \\ 0 & 1 & 0 \\ -1 & 0 & 1 \end{pmatrix}.$$

2.5 Hilbert matrices are known to be ill conditioned [30]. Let be the Hilbert matrix of order two:

$$H = \begin{pmatrix} 1 & \frac{1}{2} \\ \frac{1}{2} & \frac{1}{3} \end{pmatrix},$$

and a linear system: $H\vec{x} = \vec{b}$. Since Hilbert matrices are symmetric and positive definite, we do not need to transform them, building $M = H^T H$, but we may use directly $M = H$ in our mechanical method.

(a) Find the corresponding values of $\mu_{\pm}$, τ, α and determine the value of $|\lambda|$.
(b) Compute the estimated number of iterations needed to obtain an absolute error $\|H\vec{x}_n - \vec{b}\|$ less than 10^{-12}. Compare with Exercise 2.3.

2.6 Repeat the previous exercise, using now the Hilbert matrix of order three:

$$H = \begin{pmatrix} 1 & \frac{1}{2} & \frac{1}{3} \\ \frac{1}{2} & \frac{1}{3} & \frac{1}{4} \\ \frac{1}{3} & \frac{1}{4} & \frac{1}{5} \end{pmatrix},$$

and compare with Exercise 2.4.

2.7 Compute, for the overdamped method, the estimates $\tau_{\pm}$ for the 2×2 matrices of Exercises 2.3 and 2.5.

2.8 Repeat the previous exercise, now with the 3×3 matrices of Exercises 2.4 and 2.6.

2.9 In the study of wave propagation in one-dimensional media, for instance for chains of coupled oscillators or for partial differential equations simulated in finite differences, linear systems occur with three-diagonal symmetric matrices of the form:

$$A = \begin{pmatrix} a & b & 0 & \cdots & 0 & 0 & 0 \\ b & a & b & 0 & \cdots & 0 & 0 \\ 0 & b & a & b & 0 & \cdots & 0 \\ \vdots & \ddots & \ddots & \ddots & \ddots & \ddots & \vdots \\ 0 & \cdots & 0 & b & a & b & 0 \\ 0 & 0 & \cdots & 0 & b & a & b \\ 0 & 0 & 0 & \cdots & 0 & b & a \end{pmatrix},$$

with $a, b \in \mathbb{R}$. The eigenvalues of these matrices are:

$$a + 2|b| \cos\left(\frac{k\pi}{q+1}\right), \quad k = 1, \ldots, q, \tag{2.51}$$

q being the number of rows and columns of the matrix. The condition number of a matrix A in the matrix 2-norm $\|\cdot\|_2$ is (see Eq. (2.7.5), page 80, of [14]):

$$\kappa_2(A) = \|A\|_2 \|A^{-1}\|_2 = \frac{\sigma_{max}}{\sigma_{min}}.$$

(a) Find this condition number for A, as a function of q. Would be such a system well or ill conditioned?
(b) Find the optimal values of parameters α and τ and determine the value of $|\lambda|$. Keep in mind that, since A is symmetric we may use it as matrix M.
(c) Typical values are $a = -2$ and $b = 1$. For these values, compute the estimated number of iterations needed to obtain an absolute error $\|A\vec{x}_n - \vec{b}\|$ less than 10^{-12} as a function of q.

2.10 (a) Show that if A is a 2×2 singular matrix, the value of α given by (2.36) is zero. What value for α can you suggest in such a case?
   (b) Is α zero for any singular matrix A regardless of its dimensions?
2.11 Companion matrices appear in eigenvalue problems, in relation with characteristic polynomials

$$\lambda^q + a_{q-1}\lambda^{q-1} + \cdots + a_1\lambda + a_0.$$

They are sparse matrices with the coefficients a_k in the last column (with opposite signs) and the elements of the second diagonal, just below the main one, all equal to 1:

$$C_q = \begin{pmatrix} 0 & 0 & 0 & 0 & \cdots & -a_0 \\ 1 & 0 & 0 & 0 & \cdots & -a_1 \\ 0 & 1 & 0 & 0 & \cdots & -a_2 \\ 0 & 0 & 1 & 0 & \cdots & -a_3 \\ \vdots & \vdots & \vdots & \ddots & \ddots & \vdots \\ 0 & 0 & 0 & \cdots & 1 & -a_{q-1} \end{pmatrix}.$$

By direct computations it can be established that $C^T C$ has a first $(q-1) \times (q-1)$ block which is just the identity matrix, has values $-a_1$ to $-a_{q-1}$ as first elements of the last column and row, while the (q, q) element is $\sum_{k=0}^{q-1} a_k^2$:

$$C_q^T C_q = \begin{pmatrix} 1 & 0 & 0 & \cdots & 0 & -a_1 \\ 0 & 1 & 0 & \cdots & 0 & -a_2 \\ 0 & 0 & 1 & \cdots & 0 & -a_3 \\ \vdots & \vdots & \vdots & \ddots & \vdots & \vdots \\ 0 & 0 & 0 & \cdots & 1 & -a_{q-1} \\ -a_1 & -a_2 & -a_3 & \cdots & -a_{q-1} & \sum_{k=0}^{q-1} a_k^2 \end{pmatrix}.$$

It is easily seen that this matrix has eigenvalue $\lambda = 1$ with multiplicity $q-2$, the two other being the roots of the second order equation:

$$\lambda^2 - \left(1 + \sum_{k=0}^{q-1} a_k^2\right) \lambda + a_0^2 = 0.$$

(a) As a preliminary step, determine C_q the expression of all the eigenvalues of $C_q^T C_q$.

(b) Consider now $q = 5$, and the polynomial

$$\lambda^5 - 6\lambda^4 + 10\lambda^3 - 11\lambda + 6.$$

Compute $\mu_\pm$ and the optimal values of the parameters τ and α.

(c) Consider now $q = 4$, with arbitrary coefficients a_k. Using the previous result, determine for C_4 the values of $\mu_\pm$ and the optimal values of the parameters τ and α.

Chapter 3
Solution of Systems of Linear Equations: Numerical Simulations

3.1 Comparison with Other Similar Methods

To check the usefulness of this method, we shall compare it with the simplest and well known iterative methods: Jacobi, Gauss-Seidel, and Steepest Descent [11, 14]. We shall do this through some examples but, first, let us recall how this other methods work.

If we consider a linear system

$$A\vec{x} = \vec{b}, \tag{3.1}$$

iterative methods consist in a sequence of vectors $\{\vec{x}_n\}$ given usually by a recurrence relation of the type $\vec{x}_{n+1} = M\vec{x}_n + \vec{v}$. A method is said to be consistent with the original linear system if $\vec{x} = M\vec{x} + \vec{v} \Longrightarrow A\vec{x} = \vec{b}$, and convergent towards the solution of the linear system if it is consistent and $\lim_{n \to \infty} \vec{x}_n = \vec{x}$.

They can be understood as linear fixed point iteration methods. Convergence is then guaranteed if the spectral radius of M is strictly less than 1, that is, if the maximum of the moduli of the eigenvalues of M is less than 1. In practice, the sequence is truncated after some precision is reached.

3.1.1 Jacobi and Gauss-Seidel Methods

Sometimes Jacobi method is presented through the following decomposition of the matrix: $A = D - E - F$, where D has the diagonal part of A, $-E$ the lower-triangular part and $-F$ the upper-triangular part, the rest of the coefficients being zero. With this:

L. Vázquez and S. Jiménez, *Newtonian Nonlinear Dynamics for Complex Linear and Optimization Problems*, Nonlinear Systems and Complexity 4, DOI 10.1007/978-1-4614-5912-5_3, © Springer Science+Business Media New York 2013

$$A\vec{x} = \vec{b} \iff (D - E - F)\vec{x} = \vec{b} \iff D\vec{x} = (E + F)\vec{x} + \vec{b}$$

$$\iff \vec{x} = D^{-1}[(E + F)\vec{x} + \vec{b}], \tag{3.2}$$

supposing D invertible. We call $M_J \equiv D^{-1}(E + F)$, Jacobi's matrix and the sequence of the method is: $\vec{x}_{n+1} = M_J\vec{x}_n + D^{-1}\vec{b}$. Since D is a diagonal matrix, D^{-1} is trivially inverted. In fact, Jacobi decomposition corresponds to write an equivalent linear system, where component x_i is isolated from equation i, and then used as the recurrence. It can be expressed as

$$\vec{x}_{n+1} = M_J\vec{x}_n + D^{-1}\vec{b}$$

$$\iff (x_{n+1})_i = \frac{1}{a_{ii}}\left[-\sum_{j=1}^{i-1} a_{ij}(x_n)_j - \sum_{j=i+1}^{q} a_{ij}(x_n)_j + b_i\right], \quad i = 1, \ldots, q,$$

$$\tag{3.3}$$

where a_{ij} is the (i, j) coefficient of matrix A. The denominators correspond to the inverted diagonal elements of A.

If the computations are done keeping the order and we update the values of the components of $\vec{x}_n$ to those of $\vec{x}_{n+1}$ as soon as they are obtained, we have the Gauss-Seidel method, expressed component-wise as:

$$(x_{n+1})_i = \frac{1}{a_{ii}}\left[-\sum_{j=1}^{i-1} a_{ij}(x_{n+1})_j - \sum_{j=i+1}^{q} a_{ij}(x_n)_j + b_i\right], \quad i = 1, \ldots, q. \tag{3.4}$$

It corresponds to

$$A\vec{x} = \vec{b} \iff (D - E - F)\vec{x} = \vec{b} \iff (D - E)\vec{x} = F\vec{x} + \vec{b}$$

$$\iff \vec{x} = (D - E)^{-1}[F\vec{x} + \vec{b}], \tag{3.5}$$

and the recurrence is given by: $\vec{x}_{n+1} = M_{GS}\vec{x}_n + (D - E)^{-1}\vec{b}$, with the Gauss-Seidel matrix being $M_{GS} \equiv (D - E)^{-1}F$.

By construction both methods are consistent. But in order to ensure that either D or $D - E$ is invertible, and that the corresponding method is convergent, some preprocessing is needed [14].

3.1.2 Steepest Descent Method

When A is a symmetric, positive definite matrix, the scalar expression

$$\phi(\vec{x}) = \frac{1}{2}\vec{x}^T A \vec{x} - \vec{x}^T \vec{b} \tag{3.6}$$

has absolute minimum value $-\frac{1}{2}\vec{b}^{\mathrm{T}}A^{-1}\vec{b}$ located at $\vec{x} = A^{-1}\vec{b}$. There is no other extrema for ϕ, since A is by these hypothesis invertible. According to this, solving (3.1) and finding the location of the minimum are equivalent problems. We have used a very similar idea in our mechanical method, as we saw in the previous chapter.

A sequence is build that tends towards that point. Instead of using a dynamical system, as in our method, the idea is to follow the direction of maximal variation of ϕ. Starting from an initial vector $\vec{x}_0$, the gradient of ϕ at $\vec{x}_0$ gives us that direction. We have: $-\vec{\nabla}\phi(\vec{x}_0) = \vec{b} - A\vec{x}_0$, and we may define as an error of the method $\vec{e}_0 \equiv \vec{b} - A\vec{x}_0$. This error is not null in general (otherwise we have the exact solution), and we can, thus, look for a value α such that $\phi(\vec{x}_0 + \alpha\vec{e}_0) < \phi(\vec{x}_0)$. The value that minimizes $\phi(\vec{x}_0 + \alpha\vec{e}_0)$ corresponds to:

$$\alpha_0 = \frac{\vec{e}_0^{\mathrm{T}}\vec{e}_0}{\vec{e}_0^{\mathrm{T}}A\vec{e}_0} = \frac{1}{r(\vec{e}_0)}, \tag{3.7}$$

where $r(\vec{e}_0)$ is the Rayleigh quotient of $\vec{e}_0$ associated with matrix A [14]. We then choose $\vec{x}_1 = \vec{x}_0 + \alpha_0\vec{e}_0$, since it is a better approximation to the location of the minimum of the problem. The iteration is, at a general step n:

$$\vec{x}_{n+1} = \vec{x}_n + \frac{1}{r(\vec{e}_n)}\vec{e}_n, \quad \vec{e}_n = \vec{b} - A\vec{x}_n, \quad r(\vec{e}_n) = \frac{\vec{e}_n^{\mathrm{T}}A\vec{e}_n}{\vec{e}_n^{\mathrm{T}}\vec{e}_n}. \tag{3.8}$$

This iterative method is always convergent, but the convergence rate can be very slow as, for instance, whenever matrix A is ill-conditioned.

This method only works for A symmetric and positive definite, but it can be used for a general matrix considering instead of (3.1) the equivalent problem

$$A^{\mathrm{T}}A\vec{x} = A^{\mathrm{T}}\vec{b}, \tag{3.9}$$

but, in this case, the convergence rate is worsened.

Example 3.1.1. (Two Dimensions:). We consider a very simple case:

$$A = \begin{pmatrix} 4 & 2 \\ -1 & 3 \end{pmatrix}, \quad \vec{b} = \begin{pmatrix} 1 \\ -1 \end{pmatrix}, \quad M = \begin{pmatrix} 17 & 5 \\ 5 & 13 \end{pmatrix}. \tag{3.10}$$

In this 2×2 case, M has just two eigenvalues, μ_1 and μ_2:

$$\begin{cases} \mu_+ + \mu_- = \mu_1 + \mu_2 = \mathrm{tr}(M) = 30 \\ \mu_+ \mu_- = \mu_1 \mu_2 = \det(M) = \det(A)^2 = 14^2 \end{cases} \tag{3.11}$$

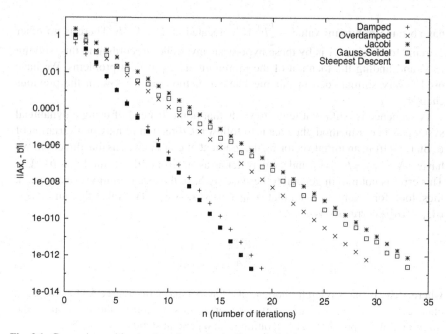

Fig. 3.1 Comparison with other iterative methods. Example 3.1.1

$$\Longrightarrow \begin{cases} \tau = \frac{2}{\sqrt{30}} \approx 0.365148372 \\ \alpha = \frac{28}{\sqrt{30}} \approx 5.112077203 \end{cases} \tag{3.12}$$

and we have $|\lambda| \approx 0.186$. If we want an error smaller than 10^{-12} we can roughly estimate the number of iterations as 17. In Figure 3.1, we compare this method with the overdamped one, Jacobi and Gauss-Seidel methods and the method of the Steepest Descent. We plot the Euclidean vector norm of the error at step n, given by $\|A\vec{x}_n - \vec{b}\|$, versus n in logarithmic scale. The alignment of the data along straight lines corresponds to the linear convergence of all these methods. The value of $\tau \approx 0.067$ for the overdamped has been chosen by trial and error to give good results. It lies close to the value that minimizes $\sqrt{(1 - \tau\mu_1)^2 + (1 - \tau\mu_2)^2}$.

Example 3.1.2. (Five Dimensions:). Let be the linear system

$$\begin{cases} -33x_1 & -9x_2 & +8x_3 & -5x_4 & -10x_5 & = & 1, \\ 2x_1 & +23x_2 & -5x_3 & +6x_4 & +10x_5 & = & -5, \\ 9x_1 & -12x_2 & +35x_3 & -7x_4 & +3x_5 & = & 7, \\ 14x_1 & +9x_2 & -8x_3 & +33x_4 & +10x_5 & = & 11, \\ -5x_1 & -15x_2 & -7x_3 & +3x_4 & +30x_5 & = & -3. \end{cases} \tag{3.13}$$

The corresponding matrices and vector are

$$A = \begin{pmatrix} -33 & -9 & 8 & -5 & -10 \\ 2 & 23 & -5 & 6 & 10 \\ 9 & -12 & 35 & -7 & 3 \\ 14 & 9 & -8 & 33 & 10 \\ -5 & -15 & -7 & 3 & 30 \end{pmatrix}, \quad \vec{b} = \begin{pmatrix} 1 \\ -5 \\ 7 \\ 11 \\ -3 \end{pmatrix}, \tag{3.14}$$

and

$$M = A^{\mathrm{T}}A = \begin{pmatrix} 1395 & 436 & -36 & 561 & 367 \\ 436 & 1060 & -574 & 519 & -76 \\ -36 & -574 & 1427 & -600 & -315 \\ 561 & 519 & -600 & 1208 & 509 \\ 367 & -76 & -315 & 509 & 1209 \end{pmatrix}. \tag{3.15}$$

Matrix A has been generated randomly but in a such a way that Jacobi and Gauss-Seidel methods can be applied to this case.

The extreme eigenvalues of M, μ_+ and μ_-, have been computed by the method described in the next two Chapters (see Example 5.4.5 on page 95). They are:

$$\mu_+ \approx 2855.965541, \quad \mu_- \approx 340.8954610, \tag{3.16}$$

and we have

$$\tau \approx 0.03537269250, \quad \alpha \approx 34.90239342. \tag{3.17}$$

For these values we have $|\lambda| \approx 0.486$. To obtain an error better than 10^{-12}, we estimate *a priori* 39 iterations. We have used the damped method with the above-mentioned values and initial vectors:

$$\vec{x}_0 = (1,0,0,0,1)^{\mathrm{T}}, \quad \vec{x}_1 = (0.98,0,0,0,0.98)^{\mathrm{T}}. \tag{3.18}$$

As we did for two dimensions, we have compared with the other standard methods: the results are plotted in Figure 3.2. We see that the estimate is optimistic and a few more iterations are needed to reach the desired accuracy. This is something to be expected in general, and it will be treated in some detail in the next Section. For the overdamped method, we have taken $\tau = 0.00063$.

As in the previous example, the damped method is one of the fastest, and we see that there is not a method which is best for both cases.

3.2 Choice of Parameter Values

In the previous Chapter we established the conditions to choose the optimal values of parameters α and τ, in order to ensure the convergence of the method under any

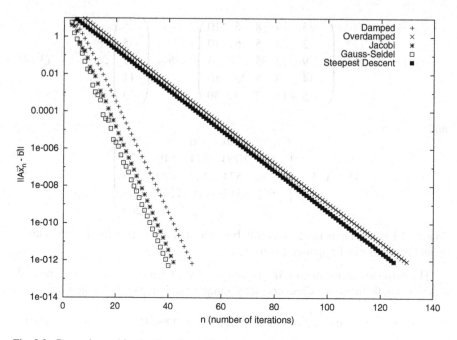

Fig. 3.2 Comparison with other iterative methods. Example 3.1.2

circumstances. We see now through an example how the damped method converges towards a solution, depending on these values.

Example 3.2.1. Let us consider the following random generated linear system with two digit coefficients in the range $[-1,1]$, given by

$$A = \begin{pmatrix} -0.08 & 0.21 & -0.56 & -0.95 & 0.86 & -0.57 & -0.01 & -0.63 \\ 0.96 & -0.74 & -0.05 & -0.84 & -0.11 & 0.61 & -0.83 & 0.79 \\ -0.45 & 0.70 & -0.18 & -0.76 & -0.11 & -0.08 & -0.41 & 0.41 \\ -0.51 & 0.22 & 0.74 & -0.93 & 0.05 & 0.73 & 0.71 & 0.36 \\ -0.24 & 0.86 & 0.43 & 0.28 & 0.97 & -0.24 & 0.09 & -0.62 \\ -0.09 & -0.30 & 0.94 & 0.10 & -0.50 & -0.48 & 0.39 & -0.82 \\ 0.90 & 0.67 & -0.11 & 0.30 & -0.52 & 0.76 & 0.94 & 0.00 \\ 0.00 & 0.54 & -0.29 & 0.15 & 0.29 & -0.28 & -0.16 & -0.60 \end{pmatrix} \qquad (3.19)$$

and $\vec{b} = (-0.97, 0.71, -0.88, 0.46, -0.91, 0.42, 0.91, -0.92)^{\mathrm{T}}$. The extreme eigenvalues of $M = A^{\mathrm{T}}A$ are:

$$\mu_+ \approx 6.408283648, \quad \mu_- \approx 0.03700095830, \qquad (3.20)$$

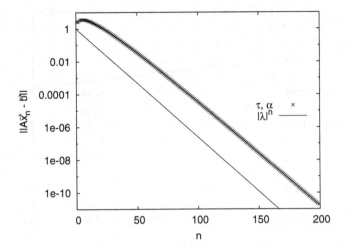

Fig. 3.3 Convergence rate, optimal values. Example 3.2.1

and the optimal values of the parameters are

$$\tau \approx 0.7877872470\,, \quad \alpha \approx 0.3836067984\,. \tag{3.21}$$

With these values we have

$$|\lambda| \approx 0.8587596275\,. \tag{3.22}$$

We have used the damped method with those values and initial vectors

$$\vec{x}_0 = (1,0,0,0,0,0,0,1)^{\mathrm{T}}, \quad \vec{x}_1 = (0.98,0,0,0,0,0,0,0.98)^{\mathrm{T}}. \tag{3.23}$$

In Figure 3.3 we plot the error of the method in semi-logarithmic scale in order to compare it with the analytical estimation $|\lambda|^n$. We see that after a transition period, the error decays basically as predicted. This transition period supposes that the *a priori* estimate is optimistic, as we saw in the previous examples.

In order to compare the convergence rate, let us also consider another set of values, not optimal:

$$\tau_1 = 0.5\,, \quad \alpha_1 = 1\,. \tag{3.24}$$

For these values, the eigenvalues of M provide three different values for $|\lambda|$:

$$|\lambda_1| \approx 0.980944258\,, \quad |\lambda_2| \approx 0.774596669\,, \quad |\lambda_3| \approx 0.611655550\,. \tag{3.25}$$

We see that

$$|\lambda| < \max_{j=1,2,3}\{|\lambda_j|\} = |\lambda_1|\,, \tag{3.26}$$

and we should expect a slower rate of convergence with the non-optimal parameters.

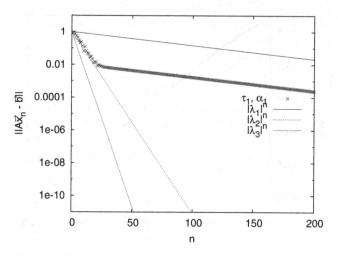

Fig. 3.4 Convergence rate, non-optimal values. Example 3.2.1

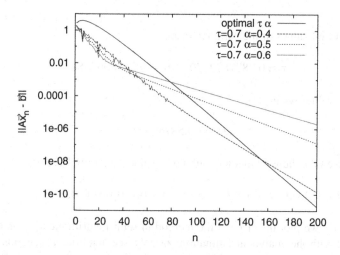

Fig. 3.5 Errors, optimal and non-optimal values. Example 3.2.1

We have plotted the error of the method for this case in Figure 3.4 and compared it with the three possible asymptotic behaviours. We have kept the same dimensions as in Figure 3.3 in order to easily compare both cases. As we see, although the second choice of parameters gives a better start, in the long run the asymptotic behaviour seems to agree with that of the maximal value of $|\lambda|$ and the result given by the optimal choice is better.

In Figure 3.5 we compare the results given by the optimal parameters and the other choices. We see that in some cases, some non-optimal values can be interesting if we do not need a good precision.

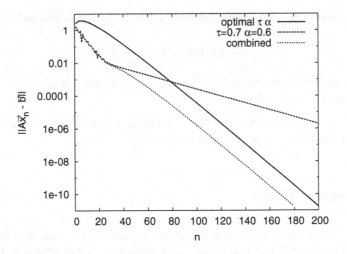

Fig. 3.6 Errors, combined iterations. Example 3.2.1

We may exploit the decay of the error for other values of the parameters in order to enhance the convergence rate, trying to avoid the initial transition period that we observe for the optimal values. Since we can always estimate the error of our numerical solution at the nth iteration step computing $\|A\vec{x}_n - \vec{b}\|$ or, better still, $\|A\vec{x}_n - \vec{b}\|/\|\vec{b}\|$, it is possible to know when the decay rate is no longer interesting and then switch to the optimal value of the parameters. We have done this for the present linear system and we represented it in Figure 3.6. We see that some 20 iterations are spared using this combined method to achieve a similar precision.

We wish to illustrate now, in an example, the behaviour of the overdamped method with respect of the choice of parameter τ and in connection with the reference value given by (2.49).

Example 3.2.2. Let us consider the same system as in previous Example 3.1.2. When we compared the different methods, we choose the value $\tau = 0.067$ for the overdamped method. From the known values of μ_+ and μ_-, the two reference values (2.49) are in this case

$$\tau_+ = \frac{1}{\mu_+} \approx 0.00035, \quad \tau_- = \frac{1}{\mu_-} \lesssim 0.0029. \tag{3.27}$$

Using τ_+, the number of iterations to achieve a relative error smaller than 10^{-12} is 214. Choosing τ_- the method does not converge. By trial and error, the best choice lies close to $\tau = 0.000618$. The number of iterations to get the same precision is 115 in that case, which more than doubles the number required by the damped method (see Figure 3.2). As we foresaw from the analysis, the overdamped method cannot compare in this sense to the damped one. And, also as predicted, the optimal value of τ lies between the two extremes.

If we perform the same numerical study as for the system in Example 3.2.1, we
have:

$$\tau_+ = \frac{1}{\mu_+} \approx 0.156, \quad \tau_- = \frac{1}{\mu_-} \lesssim 27.0. \tag{3.28}$$

The best choice seems to lie close to $\tau = 0.3$, but even with that value the
convergence rate is very slow.

From this we may conclude that the overdamped method, although much simpler
than the damped one, cannot compare in efficiency.

3.3 Singular Matrix

Equation (2.50) corresponds to a dissipative system whose solution will decay
towards a fixed point. If A is not singular, that fixed point will always be the null
vector, independently of the starting vector we choose. On the other hand, matrix
A is singular if we obtain a solution that tends to a non-null vector. To test this
possibility in an exhaustive way, we may consider as initial vector each one of the
vectors of the canonical basis, although if our system is very large in size this may
not be practical.

We illustrate all this through the following example.

Example 3.3.1. We want to test whether matrix

$$A = \begin{pmatrix} 9 & 79 & -54 & 39 & 24 & -8 & -32 \\ -2 & -26 & 18 & -12 & -8 & 2 & 10 \\ -1 & -12 & 8 & -6 & -4 & 1 & 5 \\ 6 & 76 & -54 & 38 & 24 & -6 & -30 \\ -1 & -12 & 9 & -6 & -3 & 1 & 5 \\ -4 & -47 & 32 & -22 & -14 & 5 & 18 \\ 6 & 66 & -46 & 35 & 22 & -6 & -27 \end{pmatrix} \tag{3.29}$$

is singular. In this case $\mathrm{Ker}(A) = \mathrm{span}\left\{(0,1,1,1,-1,1,1)^T\right\}$ and if we start with an
initial vector that has no projection on this subspace, the numerical solution tends to
the null vector. This is the case for the first vector of the canonical basis, but not for
any of the others: if we choose as initial vector $\vec{x}_0 = (1,0,0,0,0,0,0)^T$ the solution
tends to the null vector, while any other initial vector from the canonical basis gives
a solution proportional to $(0,1,1,1,-1,1,1)^T$. The number of iterations to reach a
given precision can be very large, as we see in Figure 3.7, but the computational time
it takes is not significant. We have used as initial vector $\vec{x}_0 = (0,1,-1,0,0,0,1)^T$.

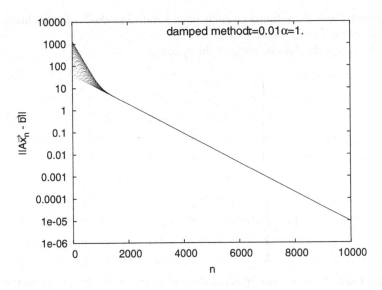

Fig. 3.7 Determining whether A is singular in Example 3.3.1

3.4 Exercises

3.1 Implement both methods, damped and overdamped, in your favorite programming language. We suggest you use $\|A\vec{x}_n - \vec{b}\|$ or $\|A\vec{x}_n - \vec{b}\|/\|\vec{b}\|$ to check the accuracy of your solution, besides considering any other control parameter you may deem useful.

3.2 Implement Jacobi, Gauss-Seidel and Steepest Descent methods in your favorite programming language. Remember that these methods may need a preprocessing of the matrix of the linear system.

3.3 Compute the lower bound for τ given by (2.39), in the previous chapter, and compare it with the actual values of τ of the Examples in this Chapter. You may include the computation of $\text{tr}(M)$ in your computer program, as a preliminary step, and then be prompted for the values of τ and α, for instance.

3.4 Solve, using the damped method, linear systems with the matrices of Exercises 2.3–2.6 and compare the actual number of iterations with the *a priori* estimates.

3.5 Repeat the previous exercise using now any of the other iterative methods and compare.

3.6 Solve, using the damped method, the two "close" ill-conditioned linear systems:

$$\begin{cases} x+2y=4, \\ 2x+4.01y=6, \end{cases} \qquad \begin{cases} x+2y=4, \\ 2.02x+4.01y=6, \end{cases}$$

Solution: $x = 404$, $y = -200$. Solution: $x = -\dfrac{404}{3}$, $y = \dfrac{208}{3}$.

Compare with the solution given by other iterative methods of your choice and also with the exact solution.

3.7 Solve, using the damped method, the systems

a)
$$\begin{cases} 2x_1 - x_2 & = 1, \\ -x_1 + 2x_2 - x_3 & = 2, \\ \quad\;\; -x_2 + 2x_3 - x_4 & = 3, \\ \quad\qquad\;\; -x_3 + 2x_4 - x_5 = 2, \\ \quad\qquad\qquad\;\; -x_4 + 2x_5 = 1, \end{cases}$$

b)
$$\begin{cases} 3x_1 + x_2 & = -2, \\ x_1 + 3x_2 + x_3 & = -1, \\ \quad\;\; +x_2 + 3x_3 + x_4 & = 0, \\ \quad\qquad\;\; +x_3 + 3x_4 + x_5 = 1, \\ \quad\qquad\qquad\;\; +x_4 + 3x_5 = 2, \end{cases}$$

(see Exercise 2.9). Keep in mind that, since the matrices are symmetric we may use them as matrix M. Use the analytical expression (2.51) to compute the optimal values of τ and α. Compute the *a priori* rate of convergence $|\lambda|$ and compare with the actual computations.

3.8 Solve, using the damped method, the system

$$\begin{cases} x_1 & -3x_5 = 2, \\ \quad x_2 & +2x_5 = 7, \\ \quad\quad x_3 & -x_5 = 8, \\ \quad\quad\quad x_4 & -2x_5 = 2, \\ -3x_1 + 2x_2 - x_3 - 2x_4 + 22x_5 = 1. \end{cases}$$

Use Exercise 2.11 and the properties presented there to compute the optimal values of τ and α. Compute the *a priori* rate of convergence $|\lambda|$ and compare with the actual computations.

3.9 Choose a linear system with a small number of dimensions (less than 15, say) of your consideration: generate it randomly or use one in connection with your fields of interests. Decide on values of τ and α, inspired by the lower bound of τ given by (2.39), and by $0 < \alpha\tau < 2$, but do not compute the optimal parameters. In case your choice does not lead to convergence try and adjust the parameters. How much *actual time* do the computations take? Determine now the optimal values for τ and α and solve the system: compare with the solution computed with your previous choice of parameters. Is there any difference in the actual time it takes the computer to give you the solution?

Comment: in many cases, and when the number of dimensions is small, the routine that outputs the solution takes much more time than the actual computations and both runs (optimal and non-optimal) take the same amount of "human" time, although the number of iterations can be quite different.

3.10 In light of the previous exercise, compare for some linear system the damped and the overdamped methods and decide whether this second method, although less efficient, can be of practical interest.

3.11 Consider the overdamped equation (2.15) and the overdamped method to test whether a given matrix A is singular or not, in a similar way as what was done for the damped equation and method.

3.12 System $A\vec{x} = \vec{b}$ with A given by (3.29) and $\vec{b} = (10, -2, 1, 4, 2, -6, 3)^\mathrm{T}$ has infinite solutions, while if $\vec{b} = (10, -1, 2, 5, 1, -5, 4)^\mathrm{T}$ there are no solutions. What are the behaviours of the different iterative methods when applied to solve them?

Chapter 4
Eigenvalue Problems

4.1 Introduction

Eigenvalue problems are at the base of many scientific and technological issues. They appear at the root of stability problems, differential equations, either ordinary or partial, Mechanics of continuous media, etc.

In the previous chapters, we proposed a method to solve systems of linear equations $A\vec{x} = \vec{b}$, by means of considering a dissipative mechanical system associated with the matrix A. This mechanical system evolves under Newton's Second Law towards the solution of the linear system. A numerical simulation was proposed then to calculate the solution in an iterative procedure.

Following a similar point of view, in this chapter we construct dynamical systems that have as sole critical points the eigenvectors of a real square matrix. In that way, all solutions of the dynamical systems, whatever initial conditions are considered, will converge to eigenvectors of the given matrix. We will see that due to the properties of the systems, the eigenvector that is approached corresponds, in principle, to the eigenvalue with minimal real part. By changing the sign of the matrix, the method converges to an eigenvector that corresponds to the eigenvalue with maximal real part. We may call these "extremal" eigenvalues and in many studies that involve a stability analysis [8, 23], their value carry important information: if the maximal real part is negative, the system is always stable, and unstable if positive. Also, the situation is not the same for the unstable case if the minimal real part is negative, since some eigenspaces are stable.

In the different sections we will develop our method: in Sect. 4.2 we present the dynamical systems and their basic properties. For the sake of readability, the corresponding proofs are presented apart, in Sect. 4.3. Finally, in Sect. 4.4, we build a simple procedure to obtain, by our method, the eigenvectors that correspond to intermediate real parts.

L. Vázquez and S. Jiménez, *Newtonian Nonlinear Dynamics for Complex Linear and Optimization Problems*, Nonlinear Systems and Complexity 4, DOI 10.1007/978-1-4614-5912-5_4, © Springer Science+Business Media New York 2013

4.2 The Dynamical Systems

4.2.1 Main Properties

Let us consider the dynamical system [19]

$$\frac{d\vec{x}}{dt} = -\frac{A}{\|\vec{x}\|^p}\vec{x} + \frac{\vec{x}^T A \vec{x}}{\|\vec{x}\|^{p+2}}\vec{x} \tag{4.1}$$

where $p \in \mathbb{R}$, $\vec{x} \in \mathbb{R}^q$ for some $q \in \mathbb{N}$ and A is a real, $q \times q$ matrix. The norm $\|\cdot\|$ is the Euclidean vector norm, as before. If A is symmetric and $p = 2$, the system is equivalent to

$$\frac{d\vec{x}}{dt} = -\vec{\nabla}U(\vec{x}), \tag{4.2}$$

where

$$U(\vec{x}) = \frac{1}{2}\frac{\vec{x}^T A \vec{x}}{\|\vec{x}\|^2}, \tag{4.3}$$

and $\vec{\nabla}$ denotes the gradient with respect to $\vec{x}$. But we are not assuming any restriction on A.

This system has the following basic properties:

1. The critical (or fixed) points are the eigenvectors of A (and conversely).
2. Conservation law:

$$\frac{d\|\vec{x}\|^2}{dt} = 0. \tag{4.4}$$

3. Existence and unicity of the solutions for any initial vector $\vec{x}(0) \neq \vec{0}$.
4. Only if A is symmetric and $p = 2$, the following dissipation law holds:

$$\frac{d}{dt}[U(\vec{x})] = -\left(\frac{d\vec{x}}{dt}\right)^2. \tag{4.5}$$

They can be established in the following way:

1. A fixed point of the dynamical system is a constant solution. Let us denote it by $\vec{v}$. We see that, because of the denominators, the null vector is not a possibility. Since $\vec{v}$ is constant, we have from (4.1):

$$-\frac{A}{\|\vec{v}\|^p}\vec{v} + \frac{\vec{v}^T A \vec{v}}{\|\vec{v}\|^{p+2}}\vec{v} = \vec{0} \iff A\vec{v} = \frac{\vec{v}^T A \vec{v}}{\|\vec{v}\|^2}\vec{v}$$

$$\iff \vec{v} \text{ is an eigenvector of } A \text{ with eigenvalue } \frac{\vec{v}^T A \vec{v}}{\|\vec{v}\|^2}. \tag{4.6}$$

Conversely, let $\vec{v}$ be an eigenvector of A with eigenvalue λ. We have:

$$A\vec{v} = \lambda\vec{v} \Longrightarrow \lambda = \frac{\vec{v}^{\mathrm{T}}A\vec{v}}{\|\vec{v}\|^2} \Longrightarrow A\vec{v} = \frac{\vec{v}^{\mathrm{T}}A\vec{v}}{\|\vec{v}\|^2}\vec{v}$$

$$\Longrightarrow \vec{x}(t) = \vec{v} \text{ is a constant solution of Eq. (4.1).} \qquad (4.7)$$

2. If we multiply (4.1) on the left by $\vec{x}^{\mathrm{T}}$, we have:

$$\vec{x}^{\mathrm{T}}\frac{d\vec{x}}{dt} = -\frac{\vec{x}^{\mathrm{T}}A\vec{x}}{\|\vec{x}\|^p} + \frac{\vec{x}^{\mathrm{T}}A\vec{x}}{\|\vec{x}\|^{p+2}}\vec{x}^{\mathrm{T}}\vec{x} = 0. \qquad (4.8)$$

On the other hand, transposing (4.1), we have:

$$\frac{d\vec{x}^{\mathrm{T}}}{dt} = -\vec{x}^{\mathrm{T}}\frac{A^{\mathrm{T}}}{\|\vec{x}\|^p} + \vec{x}^{\mathrm{T}}\frac{\vec{x}^{\mathrm{T}}A\vec{x}}{\|\vec{x}\|^{p+2}} = -\vec{x}^{\mathrm{T}}\frac{A^{\mathrm{T}}}{\|\vec{x}\|^p} + \vec{x}^{\mathrm{T}}\frac{\vec{x}^{\mathrm{T}}A^{\mathrm{T}}\vec{x}}{\|\vec{x}\|^{p+2}}, \qquad (4.9)$$

since $\vec{x}^{\mathrm{T}}A\vec{x}$ is a scalar and, thus, equal to its transposed, $\vec{x}^{\mathrm{T}}A^{\mathrm{T}}\vec{x}$. If we multiply now this expression on the right by $\vec{x}$, we obtain:

$$\frac{d\vec{x}^{\mathrm{T}}}{dt}\vec{x} = -\frac{\vec{x}^{\mathrm{T}}A^{\mathrm{T}}\vec{x}}{\|\vec{x}\|^p} + \vec{x}^{\mathrm{T}}\vec{x}\frac{\vec{x}^{\mathrm{T}}A^{\mathrm{T}}\vec{x}}{\|\vec{x}\|^{p+2}} = 0. \qquad (4.10)$$

Combining now (4.8) and (4.10), we finally have:

$$\vec{x}^{\mathrm{T}}\frac{d\vec{x}}{dt} + \frac{d\vec{x}^{\mathrm{T}}}{dt}\vec{x} = 0 \iff \frac{d(\vec{x}^{\mathrm{T}}\vec{x})}{dt} = 0 \iff \frac{d\|\vec{x}\|^2}{dt} = 0. \qquad (4.11)$$

3. The conservation law (4.4) implies that given an initial data $\vec{x}_0$, the corresponding solution lies for all times on the sphere $\|\vec{x}(t)\| = \|\vec{x}_0\|$. Since the right-hand side of (4.1) is a smooth function for $\vec{x} \neq \vec{0}$, by Chillingworth's Theorem (see, for instance, Theorem 1.0.3 in [15]), this supposes that the solution of Eq. (4.1) with initial data $\vec{x}(0) = \vec{x}_0$ always exists and is unique, provided $\vec{x}_0 \neq \vec{0}$.
4. By the chain rule derivation, we have

$$\frac{d}{dt}U(\vec{x}) = \left(\vec{\nabla}U(\vec{x})\right)^{\mathrm{T}}\frac{d\vec{x}}{dt} = -\frac{d\vec{x}^{\mathrm{T}}}{dt}\frac{d\vec{x}}{dt} = -\left(\frac{d\vec{x}}{dt}\right)^2. \qquad (4.12)$$

We have used (4.2): since we have not proved it yet, let us do it now. We suppose A symmetric and $p = 2$. We want to establish that

$$-\vec{\nabla}\left(\frac{1}{2}\frac{\vec{x}^{\mathrm{T}}A\vec{x}}{\|\vec{x}\|^2}\right) = -\frac{A}{\|\vec{x}\|^2}\vec{x} + \frac{\vec{x}^{\mathrm{T}}A\vec{x}}{\|\vec{x}\|^4}\vec{x}. \qquad (4.13)$$

By direct computations, we have:

$$\vec{\nabla}\left(\frac{1}{2}\frac{\vec{x}^T A\vec{x}}{\|\vec{x}\|^2}\right) = \frac{1}{2}\frac{\|\vec{x}\|^2\vec{\nabla}(\vec{x}^T A\vec{x}) - (\vec{x}^T A\vec{x})\vec{\nabla}(\|\vec{x}\|^2)}{\|\vec{x}\|^4}$$

$$= \frac{1}{2}\frac{\vec{\nabla}(\vec{x}^T A\vec{x})}{\|\vec{x}\|^2} - \frac{1}{2}(\vec{x}^T A\vec{x})\frac{\vec{\nabla}(\|\vec{x}\|^2)}{\|\vec{x}\|^4}$$

$$= \frac{1}{2}\frac{(A+A^T)}{\|\vec{x}\|^2}\vec{x} - \frac{1}{2}\frac{\vec{x}^T A\vec{x}}{\|\vec{x}\|^4}2\vec{x}$$

$$= \frac{A}{\|\vec{x}\|^2}\vec{x} - \frac{\vec{x}^T A\vec{x}}{\|\vec{x}\|^4}\vec{x}, \tag{4.14}$$

since A is symmetric, and, hence, (4.13) holds.

4.2.2 Properties of the Jacobian

We are going to establish the convergence of the solutions of the dynamical system towards a fixed point (i.e., towards an eigenvector of A) using the information of the linear approximation. To do this, we analyse the corresponding Jacobian. By direct computation, the Jacobian $J_p(\vec{x})$ of the dynamical system (4.1) at a given vector $\vec{x}$ is:

$$J_p(\vec{x}) = \frac{1}{\|\vec{x}\|^p}\left(\left[A - r(\vec{x})I\right]\left[pP(\vec{x}) - I\right] + P(\vec{x})\left[A + A^T - 2r(\vec{x})I\right]\right). \tag{4.15}$$

Here, I stands for the $q \times q$ identity matrix, $P(\vec{x})$ is the orthogonal projector on span$\{\vec{x}\}$ and $r(\vec{x})$ is the Rayleigh quotient at $\vec{x}$:

$$P(\vec{x}) = \frac{\vec{x}\vec{x}^+}{\|\vec{x}\|^2}, \qquad r(\vec{x}) = \frac{\vec{x}^+ A\vec{x}}{\|\vec{x}\|^2},$$

where the superscript $^+$ denotes the Hermitian conjugate which, when the vector is real, is just the transposed.

We have already used the Rayleigh quotient in (3.7), (4.1) and (4.3). Let us give here some interpretation of its meaning. Besides being the value of the corresponding eigenvalue whenever $\vec{x}$ is an eigenvector of A, $r(\vec{x})$ can be understood as the coefficient of the orthogonal projection of $A\vec{x}$ on span$\{\vec{x}\}$:

$$r(\vec{x}) = \frac{\vec{x}^+}{\|\vec{x}\|}A\frac{\vec{x}}{\|\vec{x}\|}, \tag{4.16}$$

and, in fact:

$$r(\vec{x})\vec{x} = P(\vec{x})A\vec{x}. \tag{4.17}$$

Let us consider now the linear stability of the critical points of the dynamical system. Let $\vec{u}$ be an eigenvector of A associated with the eigenvalue λ. From Lemma 1 in Sect. 4.3, we have that $J_p(\vec{u})$ is a singular matrix and that the very eigenvector $\vec{u}$ belongs to its kernel:

$$J_p(\vec{u})\vec{u} = \frac{-1}{\|\vec{u}\|^p}\left[I - P(\vec{u})\right][A - \lambda I]\vec{u} = \vec{0}. \tag{4.18}$$

Thus, we see that the Jacobian at any critical point has at least one eigenvector with zero real part. This means that, in principle, nothing can be said about the stability of the critical points from the study of the linear part [24]. If we consider A symmetric and $p = 2$, the dissipation law (4.5) would allow us to conclude that the eigenspace associated with the smallest eigenvalue is asymptotically stable. In the general case, however, this condition cannot be established in that way since the conservation law does not hold.

To deal with the general case, we have to consider that given an initial vector $\vec{x}_0$, the evolution is confined to the surface of the sphere $\|\vec{x}(t)\| = \|\vec{x}_0\|$ and, in order to check the linear stability, we must restrict ourselves to this manifold. The normal direction to the surface at $\vec{u}$ is given precisely by $\vec{u}$, thus we need to know the local behaviour around $\vec{u}$ in the orthogonal directions to $\vec{u}$. To do this, we compute all the eigenvalues and eigenvectors (either proper or generalized) of the Jacobian at $\vec{u}$, using the following result:

Theorem 4.1. *Let $\vec{u}$ be an eigenvector of A associated with the eigenvalue λ. The spectrum of A and that of $J_p(\vec{u})$ are related in the following way:*

1. *An eigenvalue λ of A corresponds to the eigenvalue 0 for $J_p(\vec{u})$*
2. *An eigenvector $\vec{w}$ associated with μ of A corresponds to*

    the eigenvector $\left[I - P(\vec{u})\right]\vec{w}$ *with eigenvalue* $\dfrac{\lambda - \mu}{\|\vec{u}\|^p}$ *for $J_p(\vec{u})$*

    and this includes the particular case where $\mu = \lambda$, the case of complex eigenvalues and eigenvectors, as well as the case of generalized eigenvectors.
3. *If the eigenspace associated with λ has algebraic multiplicity m_a and geometric multiplicity m_g such that $m_g < m_a$, then the eigenspace associated with 0 for $J_p(\vec{u})$ has algebraic multiplicity equal to m_a and geometric multiplicity equal to either m_g or $m_g + 1$. In the latter case, a generalized eigenvector of A gives rise to a proper (i.e., not generalized) eigenvector of $J_p(\vec{u})$.*
4. *If $\mu \neq \lambda$, and the eigenspace associated with μ has algebraic multiplicity m_a and geometric multiplicity m_g, then the eigenspace associated with*

$$\frac{\lambda - \mu}{\|\vec{u}\|^p}$$

for $J_p(\vec{u})$ has algebraic multiplicity equal to m_a and geometric multiplicity equal to m_g.

We prove this in Sect. 4.3.

This first result gives us all the eigenvalues of J_p, at a given eigenvector, provided we know the eigenvalues of A. We will prove now that the eigenspace associated with the eigenvalue of A with minimal real part is an attractor of the dynamical system. We start with the following stability result:

Theorem 4.2. *Let A be such that the eigenvalue with smallest real part, λ_{min}, is unique and real. Let m_a and m_g be, respectively, the algebraic and geometric multiplicities of λ_{min}. Let $\mathcal{U}_{min}$ be the set of eigenvectors associated with λ_{min} and $\bar{\mathcal{U}}_{min}$ the set of generalized eigenvectors (in case $m_a > m_g$). Then $\mathcal{U}_{min}$ is a limit set for the system, and:*

1. *Solutions inside $\mathcal{U}_{min}$ are fixed points of the system.*
2. *The components of the solutions $\vec{x}(t)$ outside $\bar{\mathcal{U}}_{min}$ decay towards an eigenvector $\vec{u} \in \mathcal{U}_{min}$ with asymptotic behaviour:*

$$\|\vec{x}(t) - \vec{u}\| \sim \exp\left[\frac{\lambda_{min} - \text{Re}(\lambda')}{\|\vec{u}\|^P} t\right]$$

   where λ' is the eigenvalue of A with real part closest to λ_{min}.
3. *The components of the solutions $\vec{x}(t)$ inside $\bar{\mathcal{U}}_{min}$ decay towards $\vec{u}$ which is the eigenvector such that $\vec{u} = (A - \lambda I)\vec{w}_1$, in the notation of Lemma 4 of Sect. 4.3, with asymptotic behaviour:*

$$\|\vec{x}(t) - \vec{u}\| \sim \frac{m_a - m_g}{t}.$$

See also Sect. 4.3 for the proof. This second result gives us the stability of the eigenspace associated with the eigenvalue of A with smaller real part, and, hence, the behaviour of our dynamical system, in the case when λ_{min} is unique and real.

We see that the convergence of a solution of the dynamical system towards the eigenvector is much faster in the case $m_a = m_g$, when λ_{min} has no generalized eigenvectors. It must be noted that if $\vec{x}_0$ has no component in $\mathcal{U}_{min}$, the corresponding solution of the dynamical system cannot reach that eigenspace. This corresponds to initial values outside $\bar{\mathcal{U}}_{min}$. In that case, we should expect convergence towards an eigenvector associated with the next smaller eigenvalue of A (provided it is unique and real). In fact, when performing numerical simulations, one would expect that small errors may give a component on $\mathcal{U}_{min}$ that gets enhanced, and the numerical solution approaches eventually $\mathcal{U}_{min}$.

As an application, we may use this dynamical system to check whether a given matrix A is singular: in that case, A^TA is a matrix with real, positive eigenvalues, having $\lambda_{min} = 0$. We will consider this in the next chapter as an alternative to the test for singular matrices of the two previous chapters.

Finally, we have a convergence result in the case where $\lambda_{\min}$ is not unique or not real, in the sense that A has eigenvalues of the form $\text{Re}_{\min} \pm i\alpha$ $(\alpha \neq 0)$:

Theorem 4.3. *Let A be real and such that there are several complex eigenvalues with smallest real part, say $\text{Re}_{\min}$. Let $\mathcal{U}$ be the direct sum of all the eigenspaces associated with those eigenvalues. Then $\mathcal{U}$ is a limit set for the system and $r(\vec{x}(t))$ converges to $\text{Re}_{\min}$ as time evolves, although $\vec{x}(t)$ does not converge necessarily to an eigenvector.*

See Sect. 4.3 for the proof. In this case, the solution of the dynamical system approaches $\mathcal{U}$, but there is not convergence towards an eigenvector. On the other hand, the Rayleigh quotient converges to the real part of the eigenvalues. Checking separately the behaviour of $\vec{x}(t)$ and of $r(\vec{x}(t))$ it is, thus, possible to identify this last case.

On the other hand, we will see in Sect. 4.4 how this case can be treated in a successful way and discriminate among the different eigenvalues that have same real part equal to the minimum one.

With these three theorems, we have established the behaviour of the dynamical system as time evolves, and how the solutions tend towards an eigenvector of A or an eigenspace of A.

4.3 Proofs

In this section, we give proofs of the three theorems. Each subsection corresponds to one of the results.

4.3.1 Spectrum of the Jacobian

In order to build the proof of Theorem 4.1, we start with some preliminary lemmas.

Lemma 1. *Let $\vec{u}$ be an eigenvector of A associated with eigenvalue λ. We then have*

$$J_p(\vec{u}) = \frac{-1}{\|\vec{u}\|^p}\left[I - P(\vec{u})\right][A - \lambda I].$$

Proof. We just consider the hypothesis and perform the computations. The result is obtained from (4.15) using the fact that $AP(\vec{u}) = \lambda P(\vec{u}) = P(\vec{u})A^{\mathsf{T}}$. □

Lemma 2. *Let $\vec{u}$ be an eigenvector of A associated with eigenvalue λ. Further, let μ be an arbitrary real value. We have*

$$\left(J_p(\vec{u}) - \frac{\lambda - \mu}{\|\vec{u}\|^p}I\right)\left[I - P(\vec{u})\right] = \frac{-1}{\|\vec{u}\|^p}\left[I - P(\vec{u})\right][A - \mu I].$$

Proof. The result is obtained by direct computations and from Lemma 1.   □

Lemma 3. *Let $\vec{u}$ and $\vec{v}$ be mutually linearly independent eigenvectors of A, associated, respectively, to eigenvalues λ and μ. Then*

$$\left[I - P(\vec{u})\right]\vec{v} \quad \text{is an eigenvector of } J_p(\vec{u}) \text{ with eigenvalue} \quad \frac{\lambda - \mu}{\|\vec{u}\|^p}.$$

Furthermore, this eigenvector of $J_p(\vec{u})$ is orthogonal to $\vec{u}$.

Proof. Using Lemma 2 we have

$$\left(J_p(\vec{u}) - \frac{\lambda - \mu}{\|\vec{u}\|^p}I\right)\left[I - P(\vec{u})\right]\vec{v} = \frac{-1}{\|\vec{u}\|^p}\left[I - P(\vec{u})\right][A - \mu I]\vec{v}.$$

This is null since $[A - \mu I]\vec{v} = \vec{0}$. On the other hand, $\vec{u}$ and $\vec{v}$ being linearly independent, we have $\left[I - P(\vec{u})\right]\vec{v} \neq \vec{0}$. Thus we prove the existence of the eigenvector and eigenvalue of $J_p(\vec{u})$.

Besides, since $I - P(\vec{u})$ is the orthogonal projector on $\text{span}\{\vec{u}\}^{\perp}$, the eigenvector is by construction orthogonal to $\vec{u}$.   □

When A is not diagonalizable, a basis of eigenvectors is not possible for the whole $\mathbb{R}^q$, and we must resort to complete with generalized eigenvectors. To do this, we must know what are these generalized eigenvectors of A. This is dealt with through the next lemmas:

Lemma 4. *Let $\vec{u}$ be an eigenvector of A associated with λ. Assume further that λ has an algebraic multiplicity m_a and a geometric multiplicity $m_g < m_a$. Let $\vec{w}_k$ be a generalized eigenvector associated with λ such that:*

$$\begin{cases} (A - \lambda I)\vec{w}_k \neq \vec{0}, \\ \quad \vdots \\ (A - \lambda I)^k \vec{w}_k \neq \vec{0}, \\ (A - \lambda I)^{k+1} \vec{w}_k = \vec{0}, \end{cases} \quad \text{with } 1 \leq k \leq m_a - m_g.$$

We then have two distinct cases:

a) *If $(A - \lambda I)\vec{w}_1 \in \text{span}\{\vec{u}\}$ then*

$$\begin{cases} J_p(\vec{u})\vec{w}_k \neq \vec{0}, \\ \quad \vdots \\ J_p^{k-1}(\vec{u})\vec{w}_k \neq \vec{0}, \\ J_p^k(\vec{u})\vec{w}_k = \vec{0}. \end{cases}$$

b) *If $(A - \lambda I)\vec{w}_1 \notin \text{span}\{\vec{u}\}$ then*

$$\begin{cases} J_p(\vec{u})\vec{w}_k \neq \vec{0}, \\ \quad \vdots \\ J_p^k(\vec{u})\vec{w}_k \neq \vec{0}, \\ J_p^{k+1}(\vec{u})\vec{w}_k = \vec{0}. \end{cases}$$

Proof. From the general properties of a defective matrix (see for instance [7]), a basis of the generalized eigenspace associated with a defective eigenvalue λ with $m_g < m_a$ can be composed out of m_g proper eigenvalues $\vec{u}_j$ with $j = 1, \ldots, m_g$ and $m_a - m_g$ generalized eigenvectors $\vec{w}_k$ with $k = 1, \ldots, m_a - m_g$, such that:

$$\forall j = 1, \ldots, m_g, \ (A - \lambda I)\vec{u}_j = \vec{0}, \qquad \forall k = 1, \ldots, m_a - m_g, \ \begin{cases} (A - \lambda I)\vec{w}_k \neq \vec{0}, \\ \vdots \\ (A - \lambda I)^k \vec{w}_k \neq \vec{0}, \\ (A - \lambda I)^{k+1} \vec{w}_k = \vec{0}. \end{cases}$$

This supposes, on the one hand, that

$$(A - \lambda I)\vec{w}_k = \vec{w}_{k-1}, \quad k = 2, \ldots, m_a - m_g,$$

and, on the other, that

$$(A - \lambda I)^k \vec{w}_k = (A - \lambda I)^{k-1} \vec{w}_{k-1} = \cdots = (A - \lambda I)\vec{w}_1, \quad k = 1, \ldots, m_g,$$

is a proper eigenvector of A.

Now, from Lemma 1 we know that for any vector $\vec{w}$

$$J_p(\vec{u})\vec{w} = \frac{-1}{\|\vec{u}\|^p} \left[I - P(\vec{u}) \right] [A - \lambda I]\vec{w}$$

and thus, to the power ℓ,

$$J_p^\ell(\vec{u})\vec{w} = \frac{(-1)^\ell}{\|\vec{u}\|^{\ell p}} \left(\left[I - P(\vec{u}) \right] [A - \lambda I] \right)^\ell \vec{w} = \frac{(-1)^\ell}{\|\vec{u}\|^{\ell p}} \left[I - P(\vec{u}) \right] [A - \lambda I]^\ell \vec{w}$$

since:

$$[A - \lambda I] \left[I - P(\vec{u}) \right] = [A - \lambda I].$$

Applying this to $\vec{w} = \vec{w}_k$ with $\ell = 1, \ldots, k+1$ proves case b). On the other hand, if

$$(A - \lambda I)\vec{w}_1 \in \text{span}\{\vec{u}\}$$

we have that

$$\left[I - P(\vec{u}) \right] [A - \lambda I]\vec{w}_1 = \vec{0}$$

and then

$$J_p(\vec{u})\vec{w}_1 = \vec{0},$$

which is the difference needed to prove case a). $\qquad\square$

Case a) always occurs if $m_g = m_a - 1$. It also occurs by chance if, among all the proper eigenvectors associated with λ, we choose our vector $\vec{u}$ belonging to $\mathrm{span}\{(A - \lambda I)\vec{w}_1\}$.

Now we study the case when the generalized eigenvectors are associated with an eigenvalue μ of A such that $\mu \neq \lambda$:

Lemma 5. *Let $\vec{u}$ be an eigenvector of A associated with λ. Let be $\mu \neq \lambda$ a defective eigenvalue of A with multiplicities $m_g < m_a$, and $\vec{w}_k$ a generalized eigenvector associated with μ, such that:*

$$\begin{cases} (A - \mu I)\vec{w}_k \neq \vec{0}, \\ \quad\vdots \\ (A - \mu I)^k \vec{w}_k \neq \vec{0}, \\ (A - \mu I)^{k+1} \vec{w}_k = \vec{0}, \end{cases} \qquad \text{with } 1 \leq k \leq m_a - m_g,$$

We then have:

$$\begin{cases} \left(J_p(\vec{u}) - \frac{\lambda - \mu}{\|\vec{u}\|^p}I\right)\left[I - P(\vec{u})\right]\vec{w}_k \neq \vec{0}, \\ \quad\vdots \\ \left(J_p(\vec{u}) - \frac{\lambda - \mu}{\|\vec{u}\|^p}I\right)^k\left[I - P(\vec{u})\right]\vec{w}_k \neq \vec{0}, \\ \left(J_p(\vec{u}) - \frac{\lambda - \mu}{\|\vec{u}\|^p}I\right)^{k+1}\left[I - P(\vec{u})\right]\vec{w}_k = \vec{0}, \end{cases} \qquad \text{with } 1 \leq k \leq m_a - m_g.$$

Proof. We use Lemma 2, repeatedly, on any vector $\vec{w}$:

$$\left(J_p(\vec{u}) - \frac{\lambda - \mu}{\|\vec{u}\|^p}I\right)^\ell \left[I - P(\vec{u})\right]\vec{w} = \frac{(-1)^\ell}{\|\vec{u}\|^{\ell p}}\left[I - P(\vec{u})\right][A - \mu I]^\ell \vec{w}$$

and apply this to $\vec{w} = \vec{w}_k$, $\ell = 1, \ldots, k + 1$. Now, contrarily to what happened in the case a) of Lemma 4, $[A - \mu I]^k \vec{w}_k$ never belongs to $\mathrm{span}\{\vec{u}\}$ since $\lambda \neq \mu$, and only when $\ell = k + 1$ can this be null.   □

Proof of Theorem 4.1 (sketch): We are now in a position to prove Theorem 4.1: the first point is proven using Lemma 3 with $\lambda = \mu$. The second point is proven using Lemma 3 in the case of proper eigenvectors, and Lemmas 4 and 5 in the case of generalized eigenvectors. It is easily seen that the complex case is also fulfilled. The third point is given directly by Lemma 4. Finally, the fourth point is proven by Lemma 5, which completes the proof.

4.3.2 Stability

As for the proof of Theorem 4.2, if the solution belongs to $\mathcal{U}_{min}$ it is an eigenvector and thus a fixed point. Even if we have several eigenvectors linearly independent (say $\{\vec{u}_i\}_{i=1}^q$, where $q = m_g$) associated with the eigenvalue, any solution of the form $\vec{x}(t) = \sum_{i=1}^q a_i(t)\vec{u}_i$ is a constant: substituting in Eq. (4.1), we have

$$\sum_{i=1}^q \dot{a}_i(t)\vec{u}_i + \frac{\lambda_{min}}{\|\vec{x}\|^p}\sum_{i=1}^q a_i(t)\vec{u}_i - \frac{\lambda_{min}}{\|\vec{x}\|^p}\sum_{i=1}^q a_i(t)\vec{u}_i = \vec{0}$$

$$\Longleftrightarrow \qquad \sum_{i=1}^q \dot{a}_i(t)\vec{u}_i = \vec{0} \Longleftrightarrow \forall i, \dot{a}_i = 0. \tag{4.19}$$

Thus, any point of eigenspace $\mathcal{U}_{min}$ is a fixed point.

We suppose thus that a general solution has components outside $\mathcal{U}_{min}$. Those can be of two types: outside $\bar{\mathcal{U}}_{min}$ and inside $\bar{\mathcal{U}}_{min} - \mathcal{U}_{min}$.

Let us start considering the first case. It is the only possibility, for instance, if $m_a = m_g$. From point 2 in Theorem 4.1, we have that all eigenvectors of $J_p(\vec{u})$ but $\vec{u}$ (either proper or generalized) belong to $\text{span}\{\vec{u}\}^\perp$. Thus the local behaviour around $\vec{u}$ is given by those other eigenvectors. Let be $\vec{u}_{min} \in \mathcal{U}_{min}$ such that it lies on the surface of the sphere $\|\vec{x}\| = \|\vec{x}_0\|$. If the geometric multiplicity m_g of λ_{min} is 1 and if we force the solutions to lie on that same sphere, it is clear that $\vec{u}_{min}$ is asymptotically stable. We only have to show that something similar is also true if $m_g > 1$. Let us denote by $\bar{\mathcal{U}}_{min}$ the set of generalized eigenvectors associated with λ_{min}. We have that, on one the hand, $J_p(\vec{u}_{min})$ has no eigenvalues with positive real part, and that all other eigenvectors associated with λ_{min} are eigenvectors of $J_p(\vec{u}_{min})$ with zero real part. On the other hand, for any other eigenvector $\vec{v}$ of A (either proper or generalized), any eigenvector of A that belongs to $\mathcal{U}_{min}$ gives rise to eigenvectors of $J_p(\vec{u})$ with negative real part. Thus any trajectory outside $\bar{\mathcal{U}}_{min}$ decays towards $\bar{\mathcal{U}}_{min}$, and its behaviour is governed by the smallest eigenvalue of $J_p(\vec{u})$ which is

$$\frac{\lambda_{min} - \lambda'}{\|\vec{u}\|^p},$$

hence the asymptotic behaviour.

We study now the second case, and consider trajectories evolving inside $\bar{\mathcal{U}}_{min}$. We will see that they tend to some proper eigenvector of λ_{min}. In order to simplify the computations we choose $\|\vec{x}(0)\| = 1$, but this poses no loss of generality. We use a notation similar to that of Lemma 4: let be $\mathcal{B} = \{\vec{u}_1, \vec{u}_2, \ldots, \vec{u}_q\}$ an orthonormal basis of $\mathcal{U}_{\min}$ and $\mathcal{B}' = \left\{ \vec{w}_1, \vec{w}_2, \ldots, \vec{w}_K \right\}$ a basis of $\bar{\mathcal{U}}_{\min} - \mathcal{U}_{\min}$ (the subspace of generalized but not proper eigenvectors of $\lambda_{\min}$) such that:

$$\begin{cases} (A - \lambda_{\min}I)\vec{w}_j = \vec{w}_{j-1}, j = 2, 3, \ldots, K, \\ (A - \lambda_{\min}I)\vec{w}_1 = \vec{u}_q. \end{cases} \tag{4.20}$$

($q = m_g$ and $K = m_a - m_g$, but we have chosen this in order to simplify the notation).
 Let us now consider a more general solution of the form

$$\vec{x}(t) = \sum_{i=1}^{q} a_i(t)\vec{u}_i + \sum_{j=1}^{K} b_j(t)\vec{w}_j. \tag{4.21}$$

Using (4.20) we have

$$A\vec{x}(t) = \lambda_{\min}\vec{x}(t) + b_1(t)\vec{u}_q + \sum_{j=1}^{K-1} b_{j+1}(t)\vec{w}_j \tag{4.22}$$

and from here

$$\vec{x}(t)^{\mathrm{T}}A\vec{x}(t) = \lambda_{\min}\|\vec{x}(t)\|^2 + b_1(t)\vec{x}(t)^{\mathrm{T}}\vec{u}_q + \sum_{j=1}^{K-1} b_{j+1}(t)\vec{x}(t)^{\mathrm{T}}\vec{w}_j. \tag{4.23}$$

Let us, for the time being, call

$$h(t) \equiv b_1(t)\vec{x}(t)^{\mathrm{T}}\vec{u}_q + \sum_{j=1}^{K-1} b_{j+1}(t)\vec{x}(t)^{\mathrm{T}}\vec{w}_j. \tag{4.24}$$

Substituting in the dynamical system we have

$$\begin{aligned} \frac{\mathrm{d}\vec{x}(t)}{\mathrm{d}t} &= -\lambda_{\min}\vec{x}(t) - b_1(t)\vec{u}_q - \sum_{j=1}^{K-1} b_{j+1}(t)\vec{w}_j + \lambda_{\min}\vec{x}(t) + h(t)\vec{x} \\ &= -b_1(t)\vec{u}_q - \sum_{j=1}^{K-1} b_{j+1}(t)\vec{w}_j + h(t)\vec{x}, \end{aligned} \tag{4.25}$$

while by direct substitution and differentiation of the solution:

$$\frac{\mathrm{d}\vec{x}(t)}{\mathrm{d}t} = \sum_{i=1}^{q} \dot{a}_i(t)\vec{u}_i + \sum_{j=1}^{K} \dot{b}_j(t)\vec{w}_j, \tag{4.26}$$

where we have, again, used the "dotted" notation to have a compact representation of time derivatives. Putting all this together, and using the fact that all the $\vec{u}$s and $\vec{w}$'s are linearly independent, we get the following system of equations for the coefficients:

$$\begin{cases} \dot{a}_i = h(t)a_i\,, \ i = 1,\ldots,q-1; \\ \dot{a}_q = h(t)a_q - b_1; \\ \dot{b}_j = h(t)b_j - b_{j+1}\,, \ j = 1,\ldots,K-1; \\ \dot{b}_K = h(t)b_K\,. \end{cases} \qquad (4.27)$$

Let us now proceed to construct the solution of system (4.27). First we will express all coefficients as functions of b_K and then solve the equation for that coefficient. We start, for instance, eliminating $h(t)$ among the equations for b_K and b_{K-1}. We have:

$$\frac{\dot{b}_{K-1} + b_K}{b_{K-1}} = \frac{\dot{b}_K}{b_K} \implies \frac{d}{dt}\left(\frac{b_{K-1}}{b_K}\right) = -1$$

$$\implies b_{K-1}(t) = (-t + \beta_{K-1})b_K(t), \qquad (4.28)$$

with the constant $\beta_{K-1} \equiv b_{K-1}(0)/b_K(0)$. We do the same with the equation for the coefficient b_{K-2} and get:

$$\frac{\dot{b}_{K-2} + b_{K-1}}{b_{K-2}} = \frac{\dot{b}_K}{b_K} \implies \frac{d}{dt}\left(\frac{b_{K-2}}{b_K}\right) = -(-t + \beta_{K-1})$$

$$\implies b_{K-2}(t) = \left(\frac{1}{2}t^2 - \beta_{K-1}t + \beta_{K-2}\right)b_K(t). \qquad (4.29)$$

We just iterate this process and get, in general:

$$b_{K-\ell}(t) = \mathcal{P}(t,\ell)b_K(t), \qquad (4.30)$$

where we have defined the family of polynomials

$$\mathcal{P}(t,\ell) \equiv \frac{(-1)^\ell}{\ell!}t^\ell + \frac{(-1)^{\ell-1}}{(\ell-1)!}\beta_{K-1}t^{\ell-1} + \frac{(-1)^{\ell-2}}{(\ell-2)!}\beta_{K-2}t^{\ell-2} + \ldots + \beta_{K-\ell}. \quad (4.31)$$

They fulfill the property:

$$\frac{d}{dt}\mathcal{P}(t,\ell) = -\mathcal{P}(t,\ell-1). \qquad (4.32)$$

It must be stressed that $\mathcal{P}(t,\ell)$ is always a polynomial of degree ℓ since the leading term cannot vanish. The constants β_{K-j} are just $b_{K-j}(0)/b_K(0)$.

The case for coefficients a_i with $i = 1,\ldots,q-1$ is simpler, yielding:

$$a_i(t) = \alpha_i b_K(t), \qquad (4.33)$$

with $\alpha_i \equiv a_i(0)/b_K(0)$. Finally, for a_q we have:

$$\frac{\dot{a}_q + b_1}{a_q} = \frac{\dot{b}_K}{b_K} \implies \frac{d}{dt}\left(\frac{a_q}{b_K}\right) = -\mathcal{P}(t, K-1)$$

$$\implies a_q(t) = \mathcal{P}(t, K)b_K(t), \tag{4.34}$$

with constant $\beta_0 \equiv a_q(0)/b_K(0)$.

We now express $h(t)$ as a function of b_K:

$$h(t) = b_1(t)\left(\sum_{i=1}^{q} a_i(t)\vec{u}_i^{\mathrm{T}} + \sum_{j=1}^{K} b_j(t)\vec{w}_j^{\mathrm{T}}\right)\vec{u}_q$$

$$+ \sum_{k=1}^{K-1} b_{k+1}(t)\left(\sum_{i=1}^{q} a_i(t)\vec{u}_i^{\mathrm{T}} + \sum_{j=1}^{K} b_j(t)\vec{w}_j^{\mathrm{T}}\right)\vec{w}_k \tag{4.35}$$

$$= b_K^2(t)\left[\mathcal{P}(t, K-1)\mathcal{P}(t, K) + \sum_{j=1}^{K}\mathcal{P}(t, K-1)\mathcal{P}(t, K-j)\vec{w}_j^{\mathrm{T}}\vec{u}_q\right.$$

$$+ \sum_{k=1}^{K-1}\sum_{i=1}^{q}\alpha_i\mathcal{P}(t, K-k-1)\vec{u}_i^{\mathrm{T}}\vec{w}_k$$

$$\left.+ \sum_{k=1}^{K-1}\sum_{j=1}^{K}\mathcal{P}(t, K-k-1)\mathcal{P}(t, K-j)\vec{w}_j^{\mathrm{T}}\vec{w}_k\right] \tag{4.36}$$

$$= b_K^2(t)\mathcal{Q}(t, 2K-1) \tag{4.37}$$

where $\mathcal{Q}(t, 2K-1)$ is just the term in square brackets in the previous expression. It is straightforward to check that it is a polynomial of degree $2K-1$ with a leading term of the form:

$$\frac{(-1)^K}{K!}t^K\frac{(-1)^{K-1}}{(K-1)!}t^{K-1} = \frac{-t^{2K-1}}{K!(K-1)!}.$$

We now solve the equation for b_K:

$$\dot{b}_K = b_K^3(t)\mathcal{Q}(t, 2K-1) \implies \frac{d}{dt}\left(\frac{-1}{2b_K^2}\right) = \mathcal{Q}(t, 2K-1)$$

$$\implies b_K(t) = \pm\left|\mathcal{R}(t, 2K)\right|^{-1/2}, \tag{4.38}$$

with $\mathcal{R}(t, 2K)$ the polynomial of degree $2K$, with leading term of the form

$$\frac{t^{2K}}{(K!)^2}, \tag{4.39}$$

such that

$$\frac{d\mathcal{R}(t,2K)}{dt} = -2\mathcal{Q}(t,2K-1), \quad \text{and} \quad \pm \left|\mathcal{R}(0,2K)\right|^{-1/2} = b_K(0).$$

We now have the solution to system (4.27):

$$\begin{cases} a_i(t) = \pm\alpha_i \left|\mathcal{R}(t,2K)\right|^{-1/2}, \ i=1,\dots,q-1; \\ a_q(t) = \pm\mathcal{P}(t,K)\left|\mathcal{R}(t,2K)\right|^{-1/2}; \\ b_j(t) = \pm\mathcal{P}(t,K-j)\left|\mathcal{R}(t,2K)\right|^{-1/2}, \ j=1,\dots,K-1; \\ b_K(t) = \pm\left|\mathcal{R}(t,2K)\right|^{-1/2}. \end{cases} \quad (4.40)$$

A few words about this solution: first of all, none of these coefficients can become singular, since $\vec{x}$ exists for all times and is bounded. Secondly, none of them can be equal to zero, unless they were zero at the initial time, and in that case remain zero for all times. This also means that the sign $(\pm)$ we should consider is just that of $b_K(0)$. Finally, due to the form of the leading terms (4.39), we see that all coefficients but $a_q(t)$ tend to zero as t goes to infinity, and that $a_q(t)$ tends to ± 1, as should be expected. The coefficient that goes to zero more slowly is $b_1(t)$, its asymptotic behaviour being

$$|b_1(t)| \sim \frac{1}{(K-1)!}t^{K-1}\left(\frac{t^{2K}}{(K!)^2}\right)^{-1/2} = \frac{K}{t} = \frac{m_a - m_g}{t}. \quad (4.41)$$

Thus we see that any trajectory inside $\mathcal{U}_{\min}$ decays towards the proper eigenvector $\vec{u}_q$. $\qquad\square$

4.3.3 Complex Case

We finally present a sketch of the proof of Theorem 4.3: in this case there is no convergence, in principle, towards an eigenvector, but as in the previous theorem, all solutions decay towards the set spanned by the eigenvectors of A associated with the eigenvalues with smallest real part. This corresponds to the space $\mathcal{U}$. On the other hand, it is easy to check that for any vector belonging to $\mathcal{U}$, the Rayleigh quotient is just $\mathrm{Re}_{\min}$.

4.4   Computation of Intermediate Eigenvalues of Matrices

At this point, we have a dynamical system that evolves, in general, towards eigenvectors corresponding to $\lambda_{\min}$. Considering matrix $-A$ instead of A, the new dynamical system gives us (changing its sign) the eigenvalue with largest real part, say $\lambda_{\max}$. We will see in the next chapter how to do this in a practical way, using a numerical method associated with (4.1).

4.4.1   General Presentation

Let us suppose, thus, that $\lambda_{\max}$ and $\lambda_{\min}$ are both known for our matrix and that both are real. We construct the following second order expression in A

$$\mathcal{P}(A) = (A - \lambda_{\max}I)(A - \lambda_{\min}I) = A^2 - (\lambda_{\max} + \lambda_{\min})A + \lambda_{\max}\lambda_{\min}I. \quad (4.42)$$

Since $\mathcal{P}(A)$ is a polynomial in A, we may obtain its Jordan canonical form with the same similarity transformation as for A. If we call J the Jordan canonical form for A, the one corresponding to $\mathcal{P}(A)$ is $\mathcal{P}(J)$. Any eigenvector of A is also an eigenvector of $\mathcal{P}(A)$. The case of generalized eigenvectors is slightly different: any generalized eigenvector of A is in principle a generalized eigenvector of $\mathcal{P}(A)$, but if the mean value

$$\lambda_{\text{mean}} \equiv \frac{\lambda_{\max} + \lambda_{\min}}{2} \quad (4.43)$$

is a defective eigenvalue of A, then the corresponding generalized eigenvectors $\vec{w}$ such that

$$\begin{cases} (A - \lambda_{\text{mean}}I)\vec{w} \neq \vec{0} \\ (A - \lambda_{\text{mean}}I)^2\vec{w} = \vec{0} \end{cases} \quad (4.44)$$

are proper eigenvectors of $\mathcal{P}(A)$ with eigenvalue $\mathcal{P}(\lambda_{\text{mean}})$. This can be checked by direct substitution:

$$A\vec{w} = \lambda_{\text{mean}}\vec{w} + \vec{u}, \quad (4.45)$$

with $\vec{u}$ a proper eigenvector of A associated with λ_{mean}, and we have:

$$\begin{aligned} \mathcal{P}(A)\vec{w} &= A^2\vec{w} - (\lambda_{\max} + \lambda_{\min})A\vec{w} + \lambda_{\max}\lambda_{\min}\vec{w} \\ &= (\lambda_{\text{mean}}^2 - (\lambda_{\max} + \lambda_{\min})\lambda_{\text{mean}} + \lambda_{\max}\lambda_{\min})\vec{w} \\ &\quad + (2\lambda_{\text{mean}} - \lambda_{\max} - \lambda_{\min})\vec{u} \\ &= \mathcal{P}(\lambda_{\text{mean}})\vec{w}. \end{aligned} \quad (4.46)$$

An example may further clarify this.

Example 4.4.1. For instance, if

$$J = \begin{pmatrix} -1 & 0 & 0 & 0 & 0 \\ 0 & 1 & 1 & 0 & 0 \\ 0 & 0 & 1 & 0 & 0 \\ 0 & 0 & 0 & 2 & 0 \\ 0 & 0 & 0 & 0 & 3 \end{pmatrix}, \text{ then } \mathcal{P}(J) = \begin{pmatrix} 0 & 0 & 0 & 0 & 0 \\ 0 & -4 & 0 & 0 & 0 \\ 0 & 0 & -4 & 0 & 0 \\ 0 & 0 & 0 & -3 & 0 \\ 0 & 0 & 0 & 0 & 0 \end{pmatrix}$$

is diagonal, even if J is not: here $\lambda_{\text{mean}} = 1$ is a defective eigenvalue, and the (2,3) component of matrix $\mathcal{P}(J)$ is 0 in this case, while if

$$J = \begin{pmatrix} -2 & 0 & 0 & 0 & 0 \\ 0 & 1 & 1 & 0 & 0 \\ 0 & 0 & 1 & 0 & 0 \\ 0 & 0 & 0 & 2 & 0 \\ 0 & 0 & 0 & 0 & 3 \end{pmatrix}, \text{ we have } \mathcal{P}(J) = \begin{pmatrix} 0 & 0 & 0 & 0 & 0 \\ 0 & -6 & 1 & 0 & 0 \\ 0 & 0 & -6 & 0 & 0 \\ 0 & 0 & 0 & -4 & 0 \\ 0 & 0 & 0 & 0 & 0 \end{pmatrix},$$

and both matrices J and $\mathcal{P}(J)$ are not diagonal, since $\lambda_{\text{mean}} = 1/2$ is not an eigenvalue.

The case of complex eigenvalues is similar: any eigenvalue of A of the form $\alpha + i\beta$ with eigenvectors of the form $\vec{u} + i\vec{v}$ is associated with the eigenvalue $\mathcal{P}(\alpha + i\beta)$ with the same eigenvectors, unless

$$\alpha = \frac{\lambda_{\text{max}} + \lambda_{\text{min}}}{2},$$

in which case $\mathcal{P}(\alpha + i\beta)$ is a real, double, eigenvalue of $\mathcal{P}(A)$ and $\vec{u}$ and $\vec{v}$ are its eigenvectors. This can also be seen by direct substitution:

$$\mathcal{P}(\alpha + i\beta) = \alpha^2 - \beta^2 - (\lambda_{\text{max}} + \lambda_{\text{min}})\alpha + \lambda_{\text{max}}\lambda_{\text{min}} + i[2\alpha - (\lambda_{\text{max}} + \lambda_{\text{min}})]\beta$$
$$= \mathcal{P}(\alpha) - \beta^2, \tag{4.47}$$

and similar for the eigenvectors. Let us illustrate this through the following example.

Example 4.4.2. Let be the real Jordan form

$$J_1 = \begin{pmatrix} 3 & -1 & 0 & 0 \\ 1 & 3 & 0 & 0 \\ 0 & 0 & 1 & 0 \\ 0 & 0 & 0 & 4 \end{pmatrix}, \text{ then } \mathcal{P}(J_1) = \begin{pmatrix} -3 & -1 & 0 & 0 \\ 1 & -3 & 0 & 0 \\ 0 & 0 & 0 & 0 \\ 0 & 0 & 0 & 0 \end{pmatrix}.$$

On the other hand, if

$$J_2 = \begin{pmatrix} 3 & -1 & 0 & 0 \\ 1 & 3 & 0 & 0 \\ 0 & 0 & 2 & 0 \\ 0 & 0 & 0 & 4 \end{pmatrix}, \text{ then } \mathcal{P}(J_2) = \begin{pmatrix} -2 & 0 & 0 & 0 \\ 0 & -2 & 0 & 0 \\ 0 & 0 & 0 & 0 \\ 0 & 0 & 0 & 0 \end{pmatrix},$$

since here $\alpha = \dfrac{\lambda_{max} + \lambda_{min}}{2}$.

If we are in the situation described by Theorem 4.3, this method separates the real eigenvalue $\lambda_{min} = \mathrm{Re}_{min}$, if it exists, from any other eigenvalue of the form $\mathrm{Re}_{min} + i\beta$. Let us suppose that $\vec{u}$ is an eigenvector associated with λ_{min} and that we have two complex conjugate eigenvalues of the form $\lambda = \lambda_{min} \pm i\beta$ with eigenvectors $\vec{v} \pm i\vec{w}$ with $\vec{v}$ and $\vec{w}$ real vectors. We have:

$$\mathcal{P}(A)(\vec{v} \pm i\vec{w}) = \mathcal{P}(\lambda_{min} \pm i\beta)(\vec{v} \pm i\vec{w}) = -\beta^2 \vec{v} \pm i\beta (\lambda_{min} - \lambda_{max})\vec{w}, \quad (4.48)$$

using (4.47) with $\alpha = \lambda_{min}$, and we see that $\vec{v} \pm i\vec{w}$ are complex eigenvectors of $\mathcal{P}(A)$ with complex eigenvalues $-\beta^2 \pm i\beta (\lambda_{min} - \lambda_{max})$. Although the original method cannot discriminate between all the eigenvalues with same minimal real part, this extension provides us with a tool to do it. Even in the case where $\lambda_{min} = \lambda_{max}$, when the complex eigenvectors and eigenvalues of $\mathcal{P}(A)$ become real, we can still separate the real eigenvalue λ_{min} from the others. Let us see it in an example.

Example 4.4.3. Let be the real Jordan form

$$J_1 = \begin{pmatrix} 2 & -1 & 0 & 0 \\ 1 & 2 & 0 & 0 \\ 0 & 0 & 2 & 0 \\ 0 & 0 & 0 & 4 \end{pmatrix}, \text{ we have: } \mathcal{P}(J_1) = \begin{pmatrix} -1 & 2 & 0 & 0 \\ -2 & -1 & 0 & 0 \\ 0 & 0 & 0 & 0 \\ 0 & 0 & 0 & 0 \end{pmatrix},$$

and, as we see, we can discriminate $\lambda_{min} = 2$ from the other two eigenvalues with same minimal real part: $2 \pm i$.

Even if $\lambda_{min} = \alpha$ and $\alpha = \dfrac{\lambda_{max} + \lambda_{min}}{2}$, which implies $\lambda_{max} = \alpha$, we can discriminate the real eigenvalues from the complex ones: if

$$J_2 = \begin{pmatrix} 2 & -1 & 0 & 0 \\ 1 & 2 & 0 & 0 \\ 0 & 0 & 2 & 0 \\ 0 & 0 & 0 & 2 \end{pmatrix}, \text{ we have: } \mathcal{P}(J_2) = \begin{pmatrix} -1 & 0 & 0 & 0 \\ 0 & -1 & 0 & 0 \\ 0 & 0 & 0 & 0 \\ 0 & 0 & 0 & 0 \end{pmatrix}.$$

Other cases, with repeated complex eigenvalues, either defective or not, can be considered and treated in a similar way. We do not present them here but leave them as exercises to the reader (See Exercises 4.7–4.13 below).

4.4.2   Statement of the method

After the previous section and its examples, we are now in a position to express the extension of the dynamical system (4.1) to deal with either complex eigenvectors or eigenvalues corresponding to intermediate real part. We reformulate (4.42) in the following way:

$$\mathcal{P}(A) = (A - \mathrm{Re}_{max} I)(A - \mathrm{Re}_{min} I)$$

$$= A^2 - (\mathrm{Re}_{max} + \mathrm{Re}_{min})A + \mathrm{Re}_{max} \, \mathrm{Re}_{min} I, \qquad (4.49)$$

where Re_{min} is the smallest real part of the eigenvalues of A and Re_{max} the largest. By construction, eigenvalues $\lambda_{min} = \mathrm{Re}_{min}$ and $\lambda_{max} = \mathrm{Re}_{max}$ correspond, if they exist, to a zero eigenvalue of $\mathcal{P}(A)$, while any real eigenvalue λ of A in $(\mathrm{Re}_{min}, \mathrm{Re}_{max})$ corresponds to a negative eigenvalue of $\mathcal{P}(A)$. Defective eigenvalues of A may correspond to non-defective for $\mathcal{P}(A)$ if

$$\lambda = \mathrm{Re}_{mean} \equiv \frac{\mathrm{Re}_{max} + \mathrm{Re}_{min}}{2}. \qquad (4.50)$$

Complex eigenvalues of A with real part $\alpha \in [\mathrm{Re}_{min}, \mathrm{Re}_{max}]$ correspond to eigenvalues, either real or complex depending on whether or not $\alpha = \mathrm{Re}_{mean}$, with, in any case, a negative real part.

The idea of the method is to simulate the dynamical system (4.1) with matrix A, then with $-A$ and extract all the possible information. Continue, replacing A by $\mathcal{P}(A)$ and simulate this new dynamical system. One by one all intermediate eigenvalues can be computed: if we identify the minimum eigenvalue of $\mathcal{P}(A)$, say μ_{min}, and an associated eigenvector $\vec{v}$, we then solve the scalar second order equation

$$\mu_{min} = z^2 - (\lambda_{max} + \lambda_{min})z + \lambda_{max}\lambda_{min}, \qquad (4.51)$$

of which one of the solutions z should correspond to λ, eigenvalue of A, with the same eigenvector $\vec{v}$. Another possibility is to calculate the corresponding Rayleigh quotient:

$$\lambda = r(\vec{v}) = \frac{\vec{v}^+ A \vec{v}}{\|\vec{v}\|^2}. \qquad (4.52)$$

We will see in the next chapter some implementations of this.

The computation of the term A^2 may be a costly operation, but it has to be evaluated just once: if an intermediate eigenvalue is obtained, a new matrix $\mathcal{P}(A)$ can be built up with just a few operations.

4.5 Exercises

4.1 (a) Are the eigenvectors of A and the fixed points of (4.1) equivalent if instead of the Euclidean vector norm, $\|\cdot\|$ represents some other norm? Use $p = 0$, for instance, to simplify the study.

(b) Does the conservation law (4.4) hold if $\|\cdot\|$ stands for a norm other than the Euclidean vector norm?

4.2 In the case of a symmetric matrix A, is it possible to have a dissipation law similar to (4.5) without having $p = 2$?

4.3 Let be the Jordan normal form

$$J = \begin{pmatrix} 2 & -3 & 0 & 0 \\ 3 & 2 & 0 & 0 \\ 0 & 0 & 2 & 1 \\ 0 & 0 & 0 & 2 \end{pmatrix}.$$

(a) List the proper eigenvector, the generalized one and the two complex.

(b) Write and solve for this case system (4.27) and check that the asymptotic behaviour of $|b_1(t)|$ corresponds to (4.41).

(c) Build J^2, use $\mathrm{Re}_{min} = \mathrm{Re}_{max} = 2$, build $\mathcal{P}(J)$ and analyse the information this provides about the eigenvectors and eigenvalues that are not real.

4.4 Let be

$$A = \begin{pmatrix} 2 & \frac{1}{2} & 0 & \frac{1}{2} \\ \frac{1}{2} & 3 & -\frac{1}{2} & -1 \\ 0 & \frac{1}{2} & 2 & \frac{1}{2} \\ \frac{1}{2} & -1 & -\frac{1}{2} & 3 \end{pmatrix}.$$

(a) Check that the eigenvectors are: $(1,0,1,0)^T$ with eigenvalue $\lambda_{min} = 2$, and $(0,1,0,-1)^T$ with eigenvalue $\lambda_{max} = 4$.
The matrix is defective: the multiplicities of eigenvalue 2 are $m_a = 3$, $m_g = 1$. We have that, for instance, a basis of $\mathcal{U}_{min} - \mathcal{U}_{min}$ is $\{(0,1,0,1)^T, (1,0,-1,0)^T\}$.

(b) Write and solve for this case system (4.27) and check that the asymptotic behaviour of $|b_1(t)|$ corresponds to (4.41).

4.5 Let be

$$
A = \begin{pmatrix}
2 & -\dfrac{1}{2} & 0 & -\dfrac{1}{2} \\
\dfrac{1}{2} & 3 & \dfrac{1}{2} & -1 \\
0 & -\dfrac{1}{2} & 2 & -\dfrac{1}{2} \\
\dfrac{1}{2} & -1 & \dfrac{1}{2} & 3
\end{pmatrix}, \quad
A^2 = \begin{pmatrix}
\dfrac{7}{2} & -2 & -\dfrac{1}{2} & -2 \\
2 & \dfrac{19}{2} & 2 & -\dfrac{13}{2} \\
-\dfrac{1}{2} & -2 & \dfrac{7}{2} & -2 \\
2 & -\dfrac{13}{2} & 2 & \dfrac{19}{2}
\end{pmatrix}.
$$

(a) Check that the eigenvalues and eigenvectors are: 2, $(-1,0,1,0)^{\mathrm{T}}$; $2\pm i$, $(\pm i,1,\pm i,1)^{\mathrm{T}}$; and 4, $(0,-1,0,1)^{\mathrm{T}}$.

(b) Compute $\mathcal{P}(A)$ and its eigenvectors and eigenvalues.

4.6 Let be

$$
A = \begin{pmatrix}
0 & 5 & -4 & -6 \\
1 & 0 & 2 & 2 \\
-3 & 13 & -9 & -14 \\
4 & -14 & 12 & 17
\end{pmatrix}, \quad
A^2 = \begin{pmatrix}
-7 & 32 & -26 & -36 \\
2 & 3 & 2 & 0 \\
-16 & 64 & -49 & -68 \\
18 & -62 & 52 & 69
\end{pmatrix}.
$$

(a) Check that the eigenvalues and eigenvectors of A are:

$$\lambda_{\max} = 3,\, (-1,-1,-2,1)^{\mathrm{T}};\, \lambda_{\min} = 1,\, (2,0,-2,1)^{\mathrm{T}};\, 2\pm i,\, (\pm i,1,\pm i,1\mp i)^{\mathrm{T}}.$$

(b) Compute $\mathcal{P}(A)$ and its eigenvectors and eigenvalues.

4.7 Let be the real Jordan form:

$$
J = \begin{pmatrix}
3 & -1 & 0 & 0 \\
1 & 3 & 0 & 0 \\
0 & 0 & 3 & -2 \\
0 & 0 & 2 & 3
\end{pmatrix},
$$

that corresponds to four complex eigenvalues with same real part. Build $\mathcal{P}(J)$ as given by (4.49) and analyse the result.

4.8 Proceed as in the previous exercise, now for the real Jordan form:

$$
J = \begin{pmatrix}
3 & -1 & 0 & 0 \\
1 & 3 & 0 & 0 \\
0 & 0 & 3 & -1 \\
0 & 0 & 1 & 3
\end{pmatrix},
$$

that corresponds to two double complex eigenvalues, non-defective.

4.9 Proceed as in the previous exercise, now for the real Jordan form:

$$J = \begin{pmatrix} 3 & -1 & 1 & 0 \\ 1 & 3 & 0 & 1 \\ 0 & 0 & 3 & -1 \\ 0 & 0 & 1 & 3 \end{pmatrix},$$

that corresponds to two double complex eigenvalues, defective.

4.10 Build $\mathcal{P}(J_1)$ and $\mathcal{P}(J_2)$, with:

$$J_1 = \begin{pmatrix} 2 & -1 & 0 & 0 & 0 & 0 \\ 1 & 2 & 0 & 0 & 0 & 0 \\ 0 & 0 & 2 & -2 & 0 & 0 \\ 0 & 0 & 2 & 2 & 0 & 0 \\ 0 & 0 & 0 & 0 & 2 & 1 \\ 0 & 0 & 0 & 0 & 0 & 2 \end{pmatrix}, \quad J_2 = \begin{pmatrix} 2 & -1 & 0 & 0 & 0 & 0 \\ 1 & 2 & 0 & 0 & 0 & 0 \\ 0 & 0 & 2 & -2 & 0 & 0 \\ 0 & 0 & 2 & 2 & 0 & 0 \\ 0 & 0 & 0 & 0 & 2 & 0 \\ 0 & 0 & 0 & 0 & 0 & 2 \end{pmatrix},$$

that correspond to four complex and a double real eigenvalue, all of them with same real part. Use $\mathrm{Re}_{\min} = \mathrm{Re}_{\max} = 2$. Compare both cases and analyse the results.

4.11 Repeat the previous exercise now with:

$$J_1 = \begin{pmatrix} 2 & -2 & 0 & 0 & 0 & 0 \\ 2 & 2 & 0 & 0 & 0 & 0 \\ 0 & 0 & 2 & -2 & 0 & 0 \\ 0 & 0 & 2 & 2 & 0 & 0 \\ 0 & 0 & 0 & 0 & 2 & 1 \\ 0 & 0 & 0 & 0 & 0 & 2 \end{pmatrix}, \quad J_2 = \begin{pmatrix} 2 & -2 & 1 & 0 & 0 & 0 \\ 2 & 2 & 0 & 1 & 0 & 0 \\ 0 & 0 & 2 & -2 & 0 & 0 \\ 0 & 0 & 2 & 2 & 0 & 0 \\ 0 & 0 & 0 & 0 & 2 & 0 \\ 0 & 0 & 0 & 0 & 0 & 2 \end{pmatrix}.$$

Discuss the differences between the results of both exercises.

4.12 Same again, now with:

$$J_1 = \begin{pmatrix} 2 & -2 & 1 & 0 & 0 & 0 \\ 2 & 2 & 0 & 1 & 0 & 0 \\ 0 & 0 & 2 & -2 & 0 & 0 \\ 0 & 0 & 2 & 2 & 0 & 0 \\ 0 & 0 & 0 & 0 & 2 & -2 \\ 0 & 0 & 0 & 0 & 2 & 2 \end{pmatrix}, \quad J_2 = \begin{pmatrix} 2 & -2 & 1 & 0 & 0 & 0 \\ 2 & 2 & 0 & 1 & 0 & 0 \\ 0 & 0 & 2 & -2 & 1 & 0 \\ 0 & 0 & 2 & 2 & 0 & 1 \\ 0 & 0 & 0 & 0 & 2 & -2 \\ 0 & 0 & 0 & 0 & 2 & 2 \end{pmatrix}.$$

4.13 Build $\mathcal{P}(J_1)$ and $\mathcal{P}(J_2)$, with:

$$J_1 = \begin{pmatrix} 1 & -1 & 0 & 0 & 0 & 0 \\ 1 & 1 & 0 & 0 & 0 & 0 \\ 0 & 0 & 3 & 1 & 0 & 0 \\ 0 & 0 & 0 & 3 & 0 & 0 \\ 0 & 0 & 0 & 0 & 5 & -1 \\ 0 & 0 & 0 & 0 & 1 & 5 \end{pmatrix}, \quad J_2 = \begin{pmatrix} 1 & -1 & 0 & 0 & 0 & 0 \\ 1 & 1 & 0 & 0 & 0 & 0 \\ 0 & 0 & 3 & 0 & 0 & 0 \\ 0 & 0 & 0 & 3 & 0 & 0 \\ 0 & 0 & 0 & 0 & 5 & -1 \\ 0 & 0 & 0 & 0 & 1 & 5 \end{pmatrix},$$

that correspond to two complex eigenvalues, and a double, intermediate real one. In both cases the intermediate real eigenvalue is equal to the mean value

$$\lambda_{mean} = \frac{Re_{max} + Re_{min}}{2}.$$

Analyse the results. Is there a significative difference between both cases, considering that in the first one the real eigenvalue is defective while not so in the second?

Chapter 5
Eigenvalue Problems: Numerical Simulations

5.1 Introduction

In this chapter a numerical scheme is proposed to simulate the evolution of the solution $\vec{x}(t)$ of dynamical system (4.1) towards an eigenvector of a given matrix, and some examples and applications are presented. The method has a linear convergence rate and we have implemented two potentially second order methods to be combined with the first one to accelerate the convergence.

Many methods have been developed to compute eigenvalue or eigenvectors of matrices. The simpler ones are the iterative Power Methods. Others rely on factorizations, such as the QR algorithm, or decomposition of the space in an orthogonal basis, as is done with Krylov subspaces [14,25]. Most of these methods are quite sophisticated and sometimes are restricted to specific problems (Hermitian or symmetric matrices, ...). The method we presented in the previous chapter and that we will implement in what follows is simple and can be used on any real matrix. It is an iterative method similar in complexity to the Power Methods. Since we will compare at some point the performance of these methods, let us now recall very briefly how these other methods work.

The simplest method is, perhaps, the Iterative or, Direct, Power Method (DPM). It corresponds to apply repeatedly matrix A to an initial vector $\vec{x}_0 \neq \vec{0}$. If there is only one eigenvalue (that may simple or not) of matrix A with largest absolute value, the iteration will converge to the associated eigenspace, since the corresponding component will be enhanced. In order to avoid either very large or very small numerical values, it is customary to normalize the resulting vector at each step, and the Euclidean vector norm is preferred in practice. The iterative process can be expressed, using an auxiliary vector, as:

$$\vec{x}_{n+1} = \frac{\vec{y}_{n+1}}{\|\vec{y}_{n+1}\|}, \quad \vec{y}_{n+1} = A\vec{x}_n. \tag{5.1}$$

L. Vázquez and S. Jiménez, *Newtonian Nonlinear Dynamics for Complex Linear and Optimization Problems*, Nonlinear Systems and Complexity 4, DOI 10.1007/978-1-4614-5912-5_5, © Springer Science+Business Media New York 2013

The method does not converge necessarily if there are two different eigenvalues of A with same maximal absolute value (real and with opposite signs), or if there are complex eigenvalues with modulus larger than the absolute value of the real ones. If the initial vector $\vec{x}_0$ has, by chance, no component on the eigenspace of this "maximal" eigenvalue, one could expect also that the convergence, if it happens, would be towards a different eigenspace. In practice, though, roundoff errors introduce the missing component and the process enhances it until convergence towards the desired eigenspace is achieved.

We may track convergence towards an eigenvalue, keeping track of a second sequence given by:

$$s_n = \vec{x}_n^T \vec{y}_{n+1} = r(\vec{x}_n), \tag{5.2}$$

since $\vec{y}_{n+1} = A\vec{x}_n$ and $\vec{x}_n$ is normalized in the Euclidean vector norm. If $\vec{x}_n$ tends to an eigenvector, s_n tends to the corresponding real eigenvalue.

If instead of considering A we use A^{-1}, supposed it is invertible, we should obtain convergence towards the eigenspace corresponding to the eigenvalue with smallest absolute value. That would be the Inverse Power Method. In practice, and since A might be singular, the matrix considered is $(A - \mu I)$ with μ some real value, and then inverted. This value μ is called the "seed" and the method is the Inverse Power Method with Seed (IPMS). The corresponding iteration can be expressed as:

$$\vec{x}_{n+1} = \frac{\vec{y}_{n+1}}{\|\vec{y}_{n+1}\|}, \quad (A - \mu I)\vec{y}_{n+1} = \vec{x}_n. \tag{5.3}$$

If convergence is achieved, it is towards an eigenvector $\vec{u}$ of matrix $(A - \mu I)$ corresponding to:

$$(A - \mu I)\vec{u} = \lambda \vec{u} \iff A\vec{u} = (\lambda + \mu)\vec{u}, \tag{5.4}$$

and, thus, we obtain an eigenvector of A with eigenvalue $(\lambda + \mu)$. It can be seen that this eigenvalue of A is the one with absolute value closest to the seed μ. Once again, convergence is not guaranteed if that eigenvalue is not real or if two real eigenvalues are at the same distance of μ but on opposite sides.

A final remark on the implementation of this last method: the linear system in (5.3) is not solved inverting the matrix. Instead, since the matrix is the same at each iteration step (μ being a fixed value), and only the right-hand side changes, a decomposition such as the LU factorization is usually performed to speed up the computations.

5.2   The Dynamical System Method

Our method that we may call the Dynamical System method (DS) consists in simulating numerically dynamical system (4.1) with a simple finite difference method, starting with an initial vector at time zero.

Using either A, $-A$ or $\mathscr{P}(A)$, we expect convergence towards the corresponding eigenvectors, and eigenvalues, as we saw in the previous chapter.

We integrate, numerically, the dynamical system starting with some initial data $\vec{x}_0$. Since our real aim is to obtain the eigenvectors of A, we want to keep the numerical method as simple as possible and we chose, for instance, a simple scheme of the form:

$$\frac{\vec{x}_{n+1} - \vec{x}_n}{\tau} = -\frac{A}{\|\vec{x}_n\|^p}\,\vec{x}_n + \frac{\vec{x}_n^{\mathrm{T}} A \vec{x}_n}{\|\vec{x}_n\|^{p+2}}\,\vec{x}_n \qquad (5.5)$$

$$\Longleftrightarrow \vec{x}_{n+1} = \vec{x}_n - \tau \frac{A}{\|\vec{x}_n\|^p}\,\vec{x}_n + \tau \frac{\vec{x}_n^{\mathrm{T}} A \vec{x}_n}{\|\vec{x}_n\|^{p+2}}\,\vec{x}_n. \qquad (5.6)$$

This has the advantage of being explicit. If we take scalar product of (5.6) with $\vec{x}_n$, we get the discrete law

$$\vec{x}_{n+1}^{\mathrm{T}} \vec{x}_n = \|\vec{x}_n\|^2 \qquad (5.7)$$

which can be seen as a discrete counterpart of the conservation of the norm (4.5). We see that this numerical scheme does not preserve the norm, but if we consider $\vec{x}_{n+1} = \vec{x}_n + \vec{\delta}_n$, we have $\vec{\delta}_n$ orthogonal to $\vec{x}_n$.

We may view the scheme as a discrete dynamical system, or an iterative method, whose fixed points are the eigenvectors of matrix A. Let $\vec{x}$ be a fixed point of (5.6), we have:

$$\vec{x} = \vec{x} - \tau \frac{A}{\|\vec{x}\|^p}\,\vec{x} + \tau \frac{\vec{x}^{\mathrm{T}} A \vec{x}}{\|\vec{x}\|^{p+2}}\,\vec{x} \iff \left(A - \frac{\vec{x}^{\mathrm{T}} A \vec{x}}{\|\vec{x}\|^2} I\right)\vec{x} = \vec{0} \qquad (5.8)$$

that corresponds to $\vec{x}$ being an eigenvector of A with eigenvalue $r(\vec{x})$. Conversely, let $\vec{u}$ be an eigenvector of A with eigenvalue λ, we have:

$$\vec{u} - \tau \frac{A}{\|\vec{u}\|^p}\,\vec{u} + \tau \frac{\vec{u}^{\mathrm{T}} A \vec{u}}{\|\vec{u}\|^{p+2}}\,\vec{u} = \vec{u} - \tau \frac{\lambda}{\|\vec{u}\|^p}\,\vec{u} + \tau \lambda \frac{\vec{u}^{\mathrm{T}} \vec{u}}{\|\vec{u}\|^{p+2}}\,\vec{u} = \vec{u}, \qquad (5.9)$$

and, thus, $\vec{x}_n = \vec{u}$ is a constant solution (a fixed point) of the iterative method.

In this way, provided the iterative method (5.6) is convergent towards one of its fixed points, we may consider the numerical scheme as a discrete method to compute eigenvectors independently and not just an approximation of the continuous dynamical system (4.1). Obviously both dynamical systems, discrete and continuous, are related as we will see in what follows.

If the numerical errors are small, we may suppose that the simulations given by the discrete method will converge to the solution of the continuous dynamical system and, thus, tend towards an eigenvector of A corresponding to the eigenvalue with minimal real part. But, on the other hand, we may study the actual convergence of the fixed point iteration on its own. For this, once again, we resort to the study of the Jacobian, which for (5.6) is:

$$J_d(\vec{x}) = I + \tau J_p(\vec{x}), \qquad (5.10)$$

where J_p is (4.15), the Jacobian of the continuous dynamical system, that we saw in the previous chapter. Let be $\vec{u}$ an eigenvector of A, i.e., a fixed point of the discrete dynamical system. If we denote by μ the eigenvalues of $J_d(\vec{u})$, the characteristic equation gives us:

$$|J_d(\vec{u}) - \mu I| = 0 \iff |I + \tau J_p(\vec{u}) - \mu I| = 0$$

$$\iff \left| J_p(\vec{u}) - \frac{\mu - 1}{\tau} I \right| = 0, \tag{5.11}$$

which means that, if we denote by γ the eigenvalues of $J_p(\vec{u})$, we have

$$\frac{\mu - 1}{\tau} = \gamma \iff \mu = 1 + \gamma\tau. \tag{5.12}$$

In principle, to ensure the convergence of the numerical scheme we need all eigenvalues of $J_d(\vec{u})$ to be of modulus less than 1. This is not possible, though, since point 1 in Theorem 4.1 tells us that $J_d(\vec{u})$ has $\gamma = 0$ as the eigenvalue associated precisely to eigenvector $\vec{u}$, and that corresponds to $\mu = 1$. But this is equivalent to $J_p(\vec{u})$ being singular and we have seen, recall Eq. (4.18) and the comments we made there, how to solve this situation: we need to know the behaviour of the system outside span$(\vec{u})$.

For instance, let us consider the case where $\lambda_{\min}$ exists and no other eigenvalue of A has Re$_{\min}$ as real part. We have, according to point 2 in Theorem 4.1, that the other eigenvalues of $J_p(\vec{u})$ are of the form

$$\gamma = \frac{\lambda_{\min} - \lambda'}{\|\vec{u}\|^p} \iff \mu = 1 + \frac{\lambda_{\min} - \lambda'}{\|\vec{u}\|^p} \tau, \tag{5.13}$$

with λ' any other eigenvalue of A. If λ' is real, for instance, $|\mu| < 1$ corresponds to

$$0 < \frac{\tau}{\|\vec{u}\|^p} < \frac{2}{\lambda' - \lambda_{\min}}, \tag{5.14}$$

with $\lambda' > \lambda_{\min}$. If λ' is the closest eigenvalue to $\lambda_{\min}$, satisfying this upper bound on τ will imply convergence. This means that the choice of τ is relevant to the convergence and, see point 2 in Theorem 4.2, also to the rate of that convergence.

We could translate the three theorems of the previous chapter to the iterative process and obtain the corresponding results, but in this case we have an advantage over the continuous dynamical system: we can perform the computations. To do that, we still have to fix some parameters in (5.6).

Numerical simulations show that for a given problem there is an optimal range of values of τ that minimizes the number of iterations required to obtain the solution with a given precision. That range depends in general not only on p but also on the choice of the initial vector, both direction and norm, although the norm that minimizes the number of iterations is usually close to 1. The case $p = 0$ is different in the sense that the number of iterations does not depend on the norm of the initial vector, but only on its direction. On the other hand, if for a given problem we fix the direction of the initial vector, but consider two different choices of p and of the initial norm, the values of τ that optimize employ the same number of iterations in both cases. We may also perform a rescaling on τ defining the new time step as $\tau/\|\vec{u}\|^p$.

From all these considerations, we have chosen to fix $p = 0$ in all the computations in this chapter. This still leaves us two aspects to be considered, of three we started with: τ and the direction of the initial vector. Of course we do not have a priori indications of which vectors are better suited: it would amount to know beforehand which the eigenvectors of A are. As it happens for the continuous dynamical system, when the initial vector has no component belonging to the eigenspace $\mathscr{U}_{\min}$, the iterative method will not converge to that eigenspace, but to some other eigenvectors, unless numerical round-off errors during the actual computations modify the situation. We have already mentioned a similar issue concerning the Direct Power Method in the introduction to this chapter. Thus, finally, we just have one parameter to consider and that is the value of τ, keeping in mind that we do not need τ to be smaller than 1, since the discrete method can be considered exact and thus without a truncation error.

It is easily seen that the value of τ such that

$$\frac{\tau}{\|\vec{u}\|^p} = \frac{1}{\lambda' - \lambda_{\min}} \tag{5.15}$$

is optimal in the sense that the component of the solution that belongs to eigenvectors associated with λ' has the fastest decay. In that case the next eigenvalue, say λ'', closest to $\lambda_{\min}$ gives the asymptotic behaviour. Once again, the optimal value of τ that minimizes the total number of iterations for a given precision is difficult to establish a priori, but it can be done minimizing the product of all non-zero eigenvalues of the Jacobian $J_d(\vec{u})$, with $\vec{u}$ is the eigenvector associated with $\lambda_{\min}$, which is the limit of the solution $\vec{x}(t)$ as t goes to infinity. For instance, (5.15) corresponds to minimizing the eigenvalue that comes from λ'. Minimizing that one and the next gives

$$\frac{\tau}{\|\vec{u}\|^p} = \frac{(\lambda' - \lambda_{\min}) + (\lambda'' - \lambda_{\min})}{2(\lambda' - \lambda_{\min})(\lambda'' - \lambda_{\min})} \tag{5.16}$$

and so on. There is not much point to this, since these values can only be deduced a posteriori.

For practical reasons, one should try a value of τ not too small: the discrete time is $t_n = \tau n$ which means that small values of τ may imply bigger values of the number of iterations n, but there is a practical way to estimate whether or not our choice is reasonable. The continuous dynamical system provides conservation of the norm, as given by Eq. (4.5). At a discrete level, though, we do not have exact conservation but an approximation given by expression (5.7). Paradoxically, we can use this loss of the conservation law to our advantage: we consider as input at step n a normalized vector $\vec{x}_n$ and compute $\vec{x}_{n+1}$ and its norm. If $\|\vec{x}_{n+1}\|$ is close to unity, our value of τ is reasonable. On the other hand, if this norm is much bigger than one, our value of τ is excessively large and should be reduced. This simple test can be performed automatically at the start and throughout all the computations, since the norm of the vector must be computed, anyway, at each iteration.

Let us illustrate the details of the method through some simple examples.

5.2.1 Some Examples in Three Dimensions

The idea of starting with examples in three dimensions is to give an insight into how the method works in situations where a graphical description of the behaviour can be presented. Although this number of dimensions is not enough to present a general case (double complex eigenvalues are excluded, for instance), we believe that these examples describe in a very visual way the essence of the method.

5.2.1.1 Choice of Parameters and Rate of Convergence

Let be

$$A = \begin{pmatrix} 6 & 0 & 0 \\ 0 & 4 & 0 \\ 0 & 0 & 1 \end{pmatrix}; \qquad \vec{x}_0 = \frac{1}{\sqrt{3}}(1,1,1)^{\mathrm{T}}. \tag{5.17}$$

We have chosen the initial vector such that it has the same component on the three eigenspaces. We have fixed the relative precision of the solution to be less than 10^{-10} in λ. We can compute the values τ_1 and τ_2 according to (5.15) and (5.16), respectively. In this case ($\lambda_{\min} = 1$, $\lambda' = 4$, $\lambda'' = 6$), they are:

$$\tau_1 = \frac{1}{3}, \qquad \tau_2 = \frac{4}{15}. \tag{5.18}$$

In Fig. 5.1 we compare the convergence for these values of τ.

We have chosen different initial vectors:

$$\vec{v}_1 = \frac{1}{\sqrt{3}}(1,1,1)^{\mathrm{T}}, \quad \vec{v}_2 = \frac{1}{\sqrt{2}}(1,0,1)^{\mathrm{T}}, \quad \vec{v}_3 = \frac{1}{\sqrt{2}}(1,1,0)^{\mathrm{T}}. \tag{5.19}$$

The first one has components along all three eigenspaces, the second one has components only along the eigenvectors associated with $\lambda_{\min}$ and λ' and the third one only along the eigenvectors associated with $\lambda_{\min}$ and λ''.

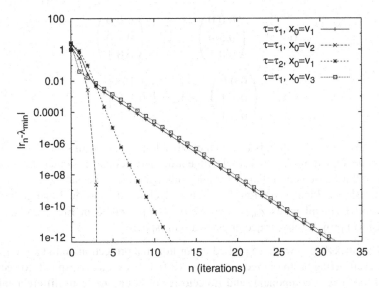

Fig. 5.1 $|r(\vec{x}_n) - \lambda_{min}|$ versus number of iterations n for matrix A with different values of τ and of the initial vector $\vec{x}_0$ according to (5.18) and (5.19). The vertical segment in the second curve means that at the following iteration step ($n = 4$) the value is exactly λ_{min}

We see that the decay with $\tau = \tau_1$ is governed by the asymptotic behaviour of the eigenvectors associated with λ'' (first and fourth curves) and that the eigenvectors associated with λ' decay superlinearly to zero (second curve). Finally τ_2 (third curve) is the optimal choice in the general case when the initial data has components along all three eigenspaces.

In this example, we may wish to compute all three eigenvalues and the corresponding eigenvectors. The minimum is computed using our method. The maximum can be obtained using the same method on the matrix $-A$, in which case we have to change the sign of the resulting eigenvalue. As to the intermediate value, we can obtain it using again the method above on A but with an initial vector with no components on $\mathcal{U}_{min}$. In this case, and once the eigenvector associated with λ_{min} is known, it is very simple. In a more general situation, it is possible to compute all eigenvalues and their eigenvectors using the ideas of Sect. 4.4.

5.2.1.2 Real Eigenvectors

Let us compare now the cases of diagonalizable versus non-diagonalizable: we will consider four cases, corresponding to matrices

$$A_1 = A = \begin{pmatrix} 6 & 0 & 0 \\ 0 & 3 & 0 \\ 0 & 0 & 1 \end{pmatrix}; \quad A_2 = \begin{pmatrix} 6 & 0 & 0 \\ 0 & 1 & 0 \\ 0 & 0 & 1 \end{pmatrix};$$

$$A_3 = \begin{pmatrix} 6 & 0 & 0 \\ 0 & 1 & 1 \\ 0 & 0 & 1 \end{pmatrix}; \quad A_4 = \begin{pmatrix} 1 & 1 & 0 \\ 0 & 1 & 1 \\ 0 & 0 & 1 \end{pmatrix}.$$

The first case corresponds to a diagonalizable problem with $\dim \mathscr{U}_{\min} = 1$. In the second one $\dim \mathscr{U}_{\min} = 2$, with algebraic and geometric dimensions equal to 2. The third case is non-diagonalizable with algebraic dimension 2 and geometric dimension 1. Finally, the fourth case has algebraic dimension 3 and geometric dimension 1. In all cases $\lambda_{\min} = 1$. We have represented in Fig. 5.2 the two-dimensional projection of trajectories on a semi-sphere.

Case 1 (matrix A_1): we have plotted the orthogonal projection on the $z = 0$ plane of the attracting basin of eigenvector $\vec{u} = (0,0,1)^{\mathrm{T}}$, which corresponds to vectors with positive z component, and trajectories of solutions from different initial values, all of them with unit norm. The arrows give the indication of movement along the solutions as time increases. The eigenvalues of the Jacobian $J_0(\vec{u})$ are -2 with eigenvector $\vec{v}_1 = (0,1,0)^{\mathrm{T}}$ and -5 with eigenvector $\vec{v}_2 = (1,0,0)^{\mathrm{T}}$. As we see, near the equilibrium point (that corresponds to the origin of the plot) the y component of the solutions decays faster than the x component, which agrees with the linear approximation. The picture is similar to that of a node in a planar system.

Case 2 (matrix A_2): $\mathscr{U}_{\min} = \mathrm{span}\{\vec{u}_1, \vec{u}_2\}$ with $\vec{u}_1 = (0,0,1)^{\mathrm{T}}$, $\vec{u}_2 = (0,1,0)^{\mathrm{T}}$, and all the points on the sphere with $x = 0$ correspond to eigenvectors of $\lambda_{\min}$. It is represented in the plot by a dotted line. Solutions tend towards an eigenvector, in fact following a geodesic on the sphere.

Case 3 (matrix A_3): $\vec{u} = (0,1,0)^{\mathrm{T}}$, the other eigenvector (both of A and of $J_0(\vec{u})$) being $\vec{v} = (1,0,0)^{\mathrm{T}}$. The third direction corresponds to a generalized eigenvector of $\lambda_{\min}$: $\vec{w} = (0,0,1)^{\mathrm{T}}$. We have plotted the orthogonal projection on the $y = 0$ plane of the solutions. Simulations show that the x component decays rapidly and, as we see in the plot, the trajectories approach the generalized eigenspace of $\lambda_{\min}$, that is $\mathrm{span}\{\vec{u}, \vec{w}\}$, and eventually they reach the equilibrium point $\vec{u}$. The general picture is similar to the one that corresponds to A_1.

The case for A_4 is clearly different. Here we have only one eigenvector: $\vec{u} = (1,0,0)^{\mathrm{T}}$, and two generalized eigenvectors $\vec{w}_1 = (0,1,0)^{\mathrm{T}}$, $\vec{w}_2 = (0,0,1)^{\mathrm{T}}$, such that $A\vec{w}_1 = \vec{u} + \vec{w}_1$ and $A\vec{w}_2 = \vec{w}_1 + \vec{w}_2$. We have projected the trajectories with positive z component on the $z = 0$ plane. The behaviour is similar to that of a parabolic sector in a planar system. As we see, near $\vec{u}$ the direction along $\vec{w}_1$ is unstable and that along $-\vec{w}_1$ stable, such that eventually all trajectories approach the equilibrium point. The behaviour on the other semi-sphere (with z negative) is similar and the solutions approach $-\vec{u}$.

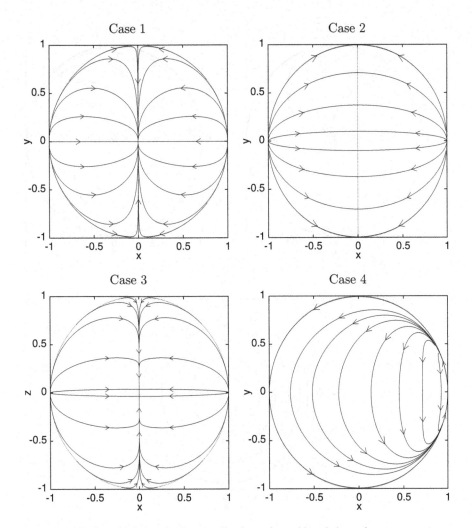

Fig. 5.2 Projection of solutions in the cases 1 to 4: matrices with real eigenvalues

5.2.1.3 Complex Eigenvectors

Although our hypothesis is that λ_{min} is real, we present in case 5 an example where the (generalized) eigenvectors of a second eigenvalue are complex. We also consider case 6 where λ_{min} is unique but non-real and there is another real eigenvalue with bigger real part. Finally we consider case 7 where λ_{min} is not unique, but there are two eigenvalues with same real part, one real and the other complex, non-real. In this two last cases, the convergence is not guaranteed since we do not fulfill the fundamental hypothesis of λ_{min} being unique and real. The matrices we are considering for these examples are:

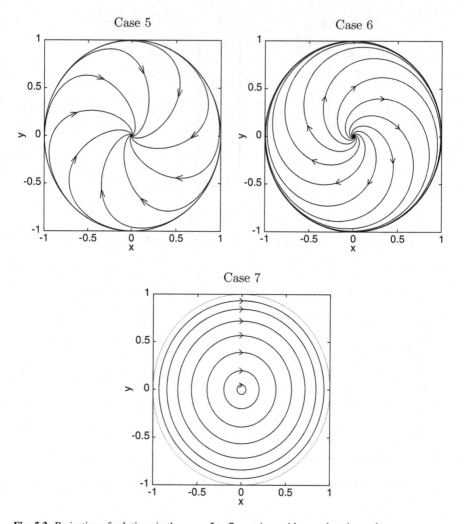

Fig. 5.3 Projection of solutions in the cases 5 to 7: matrices with complex eigenvalues

$$A_5 = \begin{pmatrix} 6 & -2 & 0 \\ 2 & 6 & 0 \\ 0 & 0 & 1 \end{pmatrix} ; \quad A_6 = \begin{pmatrix} 1 & -2 & 0 \\ 2 & 1 & 0 \\ 0 & 0 & 3 \end{pmatrix} ; \quad A_7 = \begin{pmatrix} 1 & -2 & 0 \\ 2 & 1 & 0 \\ 0 & 0 & 1 \end{pmatrix} .$$

The corresponding plots are in Fig. 5.3

In case 5, $\lambda_{\min} = 1$ associated with $(0,0,1)^T$ and we have two complex conjugate eigenvalues: $6 \pm i2$ associated with $(1,0,0)^T$ and $(0,1,0)^T$. We have plotted the projection of trajectories with $z > 0$ on the $z = 0$ plane. The eigenvector is an asymptotically stable focus.

In case 6, we have that the eigenvalues with minimal real part are $1 \pm i2$, associated with $\vec{u}_1 = (1,0,0)^T$ and $\vec{u}_2 = (0,1,0)^T$. Besides, there is a real eigenvalue, 3, associated with $(0,0,1)^T$. This last eigenvector behaves as an unstable focus and the trajectories tend to span$\{\vec{u}_1, \vec{u}_2\}$, and describe a circular motion. Although there is no convergence towards any vector, if we compute the Rayleigh quotient $r(\vec{x})$, we see that it converges towards the real part of the complex eigenvalues.

Finally, in case 7, all eigenvalues have the same real part. The complex eigenvalues are $1 \pm i2$, associated with $\vec{u}_1 = (1,0,0)^T$ and $\vec{u}_2 = (0,1,0)^T$, and the real eigenvalue is 1, associated with $(0,0,1)^T$. This eigenvector behaves as a centre and the trajectories describe a circular motion. As in the previous case, there is no convergence towards any vector, but if we compute $r(\vec{x})$, we see that it is always equal to 1.

5.2.1.4 Singular Matrices

In Sect. 2.5 we proposed a test to check whether a given real matrix A is invertible, and in Sect. 3.3 we gave an example of its application. Here we propose a different approach based on the property:

$$A \text{ is singular} \iff A^T A \text{ has eigenvalue } \lambda_{\min} = 0. \qquad (5.20)$$

What is interesting is the fact that $A^T A$ is a symmetric, positive definite matrix and, thus, our Dynamical System method should converge to an eigenvector associated with eigenvalue 0. The rate of convergence is governed by the smallest, non-zero eigenvalue λ', according to point 2 in Theorem 4.2.

Let us apply this to the matrices of Examples 2.5.1 and 3.3.1:

Example 5.2.1. Let be

$$A = \begin{pmatrix} -1 & -2 & -8 & -12 & 4 \\ -3 & 4 & 8 & 15 & -7 \\ -1 & 3 & 11 & 17 & -7 \\ 1 & -1 & -4 & -6 & 3 \\ 1 & 3 & 12 & 19 & -6 \end{pmatrix}.$$

We apply our method to $A^T A$ and obtain a very slow convergence towards a small value: in this case, although A is invertible, matrix $A^T A$ has $\lambda_{\min} \approx 0.000941$ and $\lambda' \approx 0.118$ and the method is not conclusive.

On the other hand, if we consider

$$A = \begin{pmatrix} 2 & -2 & -8 & -9 & 7 \\ -6 & 4 & 8 & 12 & -10 \\ -5 & 3 & 11 & 13 & -11 \\ 2 & -1 & -4 & -5 & 4 \\ -4 & 3 & 12 & 14 & -11 \end{pmatrix}$$

there is convergence towards 0, although it is also very slow: here $\lambda' \approx 0.0812$.

Since the IPMS does not need to build $A^T A$ in order to check whether a given matrix has a zero eigenvalue, it may be a more interesting tool to establish the singular character (see Exercise 5.12).

Example 5.2.2. We apply our method to

$$A = \begin{pmatrix} 9 & 79 & -54 & 39 & 24 & -8 & -32 \\ -2 & -26 & 18 & -12 & -8 & 2 & 10 \\ -1 & -12 & 8 & -6 & -4 & 1 & 5 \\ 6 & 76 & -54 & 38 & 24 & -6 & -30 \\ -1 & -12 & 9 & -6 & -3 & 1 & 5 \\ -4 & -47 & 32 & -22 & -14 & 5 & 18 \\ 6 & 66 & -46 & 35 & 22 & -6 & -27 \end{pmatrix} \tag{5.21}$$

and obtain slow convergence to 0. Here $\lambda' \approx 0.130$.

From the examples we conclude that solving the system seems to be a better option than computing the minimal eigenvalue.

5.2.2 Comparison with the Power Methods

We will now compare the performance of our method and that of the Direct Power Method (DPM) and the Inverse Power Method with Seed (IPMS) [7] that we presented in the introduction to this chapter.

We define different matrices with a specific spectrum and use them as test data. Consider matrices

$$M_1 = \begin{pmatrix} -6 & 6 & -4 & 4 & 1 \\ -23 & 21 & -14 & 14 & 5 \\ -23 & 20 & -14 & 15 & 6 \\ 22 & -18 & 13 & -12 & -5 \\ -68 & 56 & -42 & 42 & 17 \end{pmatrix}, M_2 = \begin{pmatrix} -7 & 8 & -5 & 6 & 2 \\ -24 & 23 & -15 & 16 & 6 \\ -23 & 20 & -14 & 15 & 6 \\ 13 & -10 & 8 & -8 & -4 \\ -46 & 37 & -30 & 33 & 15 \end{pmatrix}.$$

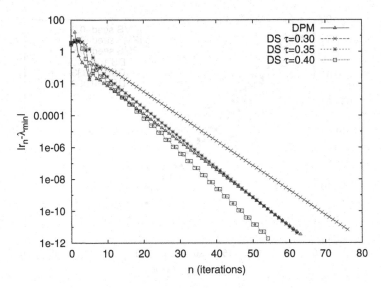

Fig. 5.4 Comparison of $|r(\vec{x}_n) - \lambda_{\max}|$ versus number of iterations n for matrix M_1 with the Power Method and the dynamical system with three different values of τ

They have canonical Jordan forms Λ_i, such that $P\Lambda_i P^{-1} = M_i$ with:

$$\Lambda_1 = \begin{pmatrix} -1 & 0 & 0 & 0 & 0 \\ 0 & 1 & 1 & 0 & 0 \\ 0 & 0 & 1 & 0 & 0 \\ 0 & 0 & 0 & 2 & 0 \\ 0 & 0 & 0 & 0 & 3 \end{pmatrix}, \quad \Lambda_2 = \begin{pmatrix} 1 & 1 & 0 & 0 & 0 \\ 0 & 1 & 1 & 0 & 0 \\ 0 & 0 & 1 & 0 & 0 \\ 0 & 0 & 0 & 3 & 1 \\ 0 & 0 & 0 & 0 & 3 \end{pmatrix},$$

$$P = \begin{pmatrix} 1 & 0 & 1 & 1 & 0 \\ 1 & 0 & 0 & 1 & 1 \\ 0 & 1 & -2 & 0 & 1 \\ -1 & 1 & 0 & 1 & -1 \\ 3 & 0 & -1 & -2 & 2 \end{pmatrix}.$$

Matrix P has been chosen assigning arbitrarily the values $0, \pm 1, \pm 2$ to its elements (an element has been changed to 3 so that the resulting matrices M_i have all their elements as integer values, for the sake of simplicity). In this way, the eigenvectors are not mutually orthogonal.

In Fig. 5.4 we compare the results of simulating the associated dynamical system (DS) with different values of τ and of the DPM for matrix M_1 in order to obtain $\lambda_{\max} \equiv 3$. To have DS converging towards the maximum eigenvalue, we have taken $-M_1$. In all computations the initial vector is $(0, 0, 1, 1, 1)^{\mathrm{T}}$, normalized in

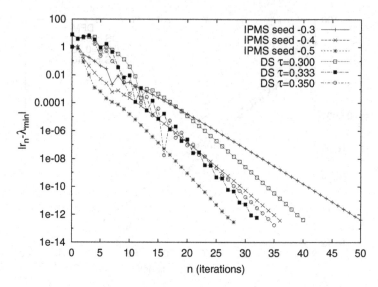

Fig. 5.5 Comparison of $|r(\vec{x}_n) - \lambda_{\min}|$ versus number of iterations n for matrix M_1 with the Inverse Power Method with three seeds and the dynamical system with three different values of τ

the Euclidean vector norm. As we see, the number of iterations is similar for both methods, provided we chose a reasonable value of τ for the DS.

In Fig. 5.5 we compare the results of simulating the DS with different values of τ and of the IPMS for matrix M_1 in order to obtain $\lambda_{\min} = -1$. We cannot start the IPMS with seed 0, since 1 is also an eigenvalue and the method does not converge. As we see, the number of iterations is similar for both methods, provided we chose a reasonable value of τ for the DS and of the seed for the IPMS.

In Fig. 5.6 we compare the results of simulating the DS with different values of τ and of the DPM for matrix M_2 in order to obtain $\lambda_{\max} \equiv 3$. In order to have DS converging towards the maximum eigenvalue, we have taken $-M_2$. In all computations the initial vector is $(0, 0, 1, 1, 1)^{\mathrm{T}}$, normalized. As we see, the precision of both methods is similar for a number of iterations fixed. We are in the case of generalized eigenvectors where the decay towards the eigenvector is not exponential.

In Fig. 5.7 we compare the results of simulating the dynamical system (DS) with different values of τ and of the IPMS for matrix M_2 in order to obtain $\lambda_{\min} = 1$. We start the IPMS with seed 0. The precision of both methods is similar for a number of iterations fixed.

As we see, the different kinds of behaviour allow in practice to know whether the associated eigenvalue is real and unique or if there are others with the same real part. We also can deduce whether there are generalized eigenvectors.

From the numerical point of view, the performance of our method is similar to that of the Power Methods, with the difference of the conditions of applicability: this method can be used in situations where the Power Methods do not converge because there are two eigenvalues with same absolute value or equally distant from the seed.

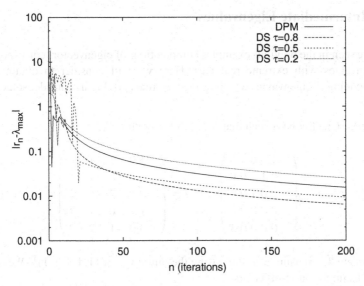

Fig. 5.6 Comparison of $|r(\vec{x}_n) - \lambda_{\max}|$ versus number of iterations n for matrix M_2 with the Power Method and the dynamical system with three different values of τ. Due to the big number of iterations, the data are represented by lines rather that points

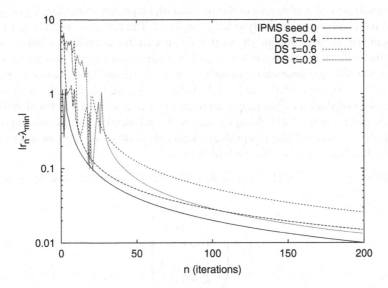

Fig. 5.7 Comparison of $|r(\vec{x}_n) - \lambda_{\min}|$ versus number of iterations n for matrix M_2 with the Inverse Power Method and the dynamical system with three different values of τ

5.3   Intermediate Eigenvalues

We have seen in the previous examples computations of eigenvectors corresponding to eigenvalues with extreme real part. Here we will consider to obtain those corresponding to eigenvalues with intermediate real part. Let us consider some 5×5 examples:

Example 5.3.1. Let be A such that $A = QJQ^{-1}$, with

$$
J = \begin{pmatrix} -1 & 0 & 0 & 0 & 0 \\ 0 & 1 & 1 & 0 & 0 \\ 0 & 0 & 1 & 0 & 0 \\ 0 & 0 & 0 & 2 & 0 \\ 0 & 0 & 0 & 0 & 3 \end{pmatrix}, \quad Q = \begin{pmatrix} 1 & 0 & 1 & 1 & 0 \\ 1 & 0 & 0 & 1 & 1 \\ 0 & 1 & -2 & 0 & 1 \\ -1 & 1 & 0 & 1 & -1 \\ 3 & 0 & -1 & -2 & 2 \end{pmatrix}. \tag{5.22}
$$

We simulate the dynamical system using the initial vector $(1,1,1,1,1)^{\mathrm{T}}$. We obtain for (4.51), up to round-off errors:

$$
\mu_{\min} = -4, \quad \vec{v} = \frac{1}{\sqrt{6}}(1,0,0,2,-1)^{\mathrm{T}}, \quad z = 1 \text{ (double)}, \quad r(\vec{v}) = \frac{4}{3}.
$$

The fact that z is a double root of the second order equation shows that eigenvalue 1 of A has algebraic multiplicity at least equal to 2. The fact that z and the Rayleigh quotient $r(\vec{v})$ do not coincide shows that $\vec{v}$ is not an eigenvector of A but a linear combination of a proper eigenvalue $\vec{u}$ and of a generalized one $\vec{w}$ associated with $\lambda = 1$ and thus the geometric multiplicity is not maximal. In this case we have: $\vec{u} = (0,0,1,1,0)^{\mathrm{T}}, \vec{w} = (1,0,-2,0,-1)^{\mathrm{T}}$. As we see, we are able from the analysis of the results to deduce the main properties of the eigenvalues. Once $\lambda = 1$ is identified, we can build a new $\mathscr{P}(A)$ changing $\lambda_{\min}$ to 1, and identify the last intermediate eigenvalue 2. With all the eigenvalues identified, in this case we can assign the proper algebraic and geometric multiplicities.

Example 5.3.2. We consider A such that $A = QJQ^{-1}$, with the same Q as in the previous example and

$$
J = \begin{pmatrix} -2 & 0 & 0 & 0 & 0 \\ 0 & 1 & 0 & 0 & 0 \\ 0 & 0 & 1 & 1 & 0 \\ 0 & 0 & 0 & 1 & 0 \\ 0 & 0 & 0 & 0 & 2 \end{pmatrix}. \tag{5.23}
$$

We simulate the dynamical system using the same initial data as before and we obtain:

$$
\mu_{\min} = -3, \quad \vec{v} = \frac{1}{\sqrt{6}}(1,0,0,2,-1)^{\mathrm{T}}, \quad z = \pm 1, \quad r(\vec{v}) = 1.
$$

The results show that in this case the eigenvalue is 1 with eigenvector $\vec{v}$. If we compute now changing λ_{min} by 1, we get:

$$\mu_{min} = 0, \quad \vec{v} = \frac{1}{\sqrt{5}}(1,1,1,1,1)^{\mathrm{T}}, \quad z_+ = 2, \ z_- = 1, \quad r(\vec{v}) = 1.6$$

and we conclude that there are no more eigenvalues of A besides the ones we already have.

Example 5.3.3. We consider now $A = QJQ^{-1}$, with the same Q as before but now

$$J = \begin{pmatrix} -1 & 0 & 0 & 0 & 0 \\ 0 & 1 & -2 & 0 & 0 \\ 0 & 2 & 1 & 0 & 0 \\ 0 & 0 & 0 & 2 & 0 \\ 0 & 0 & 0 & 0 & 3 \end{pmatrix}. \tag{5.24}$$

The results, starting the simulations with the same initial vector as before, are:

$$\mu_{min} = -8, \quad \vec{v} = \frac{1}{\sqrt{6}}(1,0,0,2,-1)^{\mathrm{T}}, \quad z = 1 \pm \mathrm{i}2, \quad r(v) = \frac{5}{3}.$$

We identify $1 \pm \mathrm{i}2$ as the complex eigenvalues of A. Vector $\vec{v}$ is a linear combination of the real and imaginary part of the eigenvectors.

When searching for eigenvectors corresponding to intermediate eigenvalues, we may use the information that we already have about known eigenspaces, for instance those associated with λ_{min} and λ_{max}, to choose an initial vector with components along other directions. It is not necessary, though, to perform a systematical orthogonal decomposition of the space.

5.4 Enhancing the Convergence Rate

From Theorem 4.2 we may expect, in the best of scenarios, an exponential decay of the error. This corresponds to a linear convergence of the method. It is, thus, interesting to analyse whether a better convergence rate can be achieved.

The next step would be to obtain superlinear or quadratic convergence. The easiest way to build a method that is in general quadratic is simply to solve an associated equation by Newton's method.

5.4.1   A First Attempt

For instance, as a first attempt, we can seek the zeros of the right-hand side of (4.1). We may consider, thus,

$$-\frac{A}{\|\vec{x}\|^p}\,\vec{x} + \frac{\vec{x}^T A \vec{x}}{\|\vec{x}\|^{p+2}}\,\vec{x} = \vec{0} \tag{5.25}$$

or, equivalently,

$$\left[A - r(\vec{x})I\right]\vec{x} = \vec{0}. \tag{5.26}$$

Newton's method for a vector equation of the form $\vec{f}(\vec{x}) = \vec{0}$ corresponds to the iteration given by

$$\vec{x}_{n+1} = \vec{x}_n - [Df(\vec{x}_n)]^{-1}\vec{f}(\vec{x}_n), \tag{5.27}$$

where $Df(\vec{x})$ is the Jacobian matrix of $\vec{f}$ computed at vector $\vec{x}$ [28]. A necessary condition for a quadratic convergence towards a solution $\vec{u}$ is $Df(\vec{u})$ not being singular. If we consider the left-hand side of (5.26) as $\vec{f}$, the Jacobian Df is simply $-J_p$, setting $p = 0$ in (4.15), and we have

$$\begin{aligned} Df(\vec{x}) &= \left[A - r(\vec{x})I\right] - P(\vec{x})\left[A + A^T - 2r(\vec{x})I\right] \\ &= [I - P(\vec{x})]\,[A - r(\vec{x})I] - P(\vec{x})\,[A - r(\vec{x})I]^T. \end{aligned} \tag{5.28}$$

But we saw already in Lemma 1 of Sect. 4.3 that $J_p(\vec{u})$ is singular for any eigenvector $\vec{u}$, and $Df(\vec{u})$ inherits this property. In fact, for any eigenvector $\vec{u}$ of A, we have

$$Df(\vec{u}) = [I - P(\vec{u})]\,[A - \lambda I], \tag{5.29}$$

and, clearly, the determinant $|Df(\vec{u})|$ is a double zero, since both factors in (5.29) correspond to a singular matrix. We cannot have quadratic convergence of Newton's method applied to our chosen $\vec{f}$.

5.4.2   A Better Choice

There is a condition we have not exploited, and it is the fact that the norm of the vector is non-zero. We may add one extra condition and one extra variable and consider the set of equations:

$$\vec{f}(\vec{x}, \eta) = \vec{0}_{q+1} \iff \begin{cases} (A - \eta I)\vec{x} = \vec{0}_q, \\ \|\vec{x}\|^2 - 1 = 0, \end{cases} \tag{5.30}$$

where $\vec{f}$ is now a vector function with $q + 1$ components, and $\vec{0}_q$ and $\vec{0}_{q+1}$ are the null vectors of $\mathbb{R}^q$ and $\mathbb{R}^{q+1}$, respectively. We present here this method for the case of matrices, but the general setting for eigenvalues and eigenvectors of linear operators in Banach spaces can be found in [2]. The corresponding Jacobian in the $q+1$ variables is given in block form by

$$Df(\vec{x}, \eta) = \left(\begin{array}{c|c} A - \eta I & -\vec{x} \\ \hline 2\vec{x}^{\mathrm{T}} & 0 \end{array}\right), \tag{5.31}$$

where the first block in the diagonal has $q \times q$ dimensions and the second one is just the scalar zero (thus, 1×1). The other two blocks, off the diagonal, are a column and a row of dimension q.

Let be J the canonical Jordan form of matrix A, and Q the corresponding change of basis matrix such that

$$Q^{-1}AQ = J, \tag{5.32}$$

and such that $Q^{-1}\vec{u}$ is the first vector in the Jordan basis. Let M be the $(q+1) \times (q+1)$ non-singular matrix whose block representation is

$$M = \left(\begin{array}{c|c} Q & \vec{0}_q \\ \hline \vec{0}_q^{\mathrm{T}} & 1 \end{array}\right). \tag{5.33}$$

When computing Df at a solution $(\vec{u}, \lambda)$ we have, on the one hand,

$$|Df(\vec{u}, \lambda)| = |M^{-1}Df(\vec{u}, \lambda)M|, \tag{5.34}$$

and on the other:

$$M^{-1}Df(\vec{u}, \lambda)M = \left(\begin{array}{c|c} Q^{-1} & \vec{0}_q \\ \hline \vec{0}_q^{\mathrm{T}} & 1 \end{array}\right) \left(\begin{array}{c|c} A - \lambda I & -\vec{u} \\ \hline 2\vec{u}^{\mathrm{T}} & 0 \end{array}\right) \left(\begin{array}{c|c} Q & \vec{0}_q \\ \hline \vec{0}_q^{\mathrm{T}} & 1 \end{array}\right)$$

$$= \left(\begin{array}{c|c} J - \lambda I & -Q^{-1}\vec{u} \\ \hline 2\vec{u}^{\mathrm{T}}Q & 0 \end{array}\right). \tag{5.35}$$

Our choice in ordering the eigenvalues and the eigenvectors in the Jordan form and basis is such that

$$Q^{-1}\vec{u} = (1, 0, \dots, 0)^{\mathrm{T}}, \tag{5.36}$$

and, if we denote by λ_1 through λ_q the q roots of the characteristic polynomial of A (not all necessarily different) as they appear in the diagonal of matrix J, we have $\lambda_1 = \lambda$. Finally, let us define $\vec{v} = Q^{\mathrm{T}}\vec{u}$. With all of the above, we have:

$$M^{-1}Df(\vec{u},\lambda)M = \begin{pmatrix} \lambda_1 - \lambda & \delta_1 & 0 & \cdots & 0 & -1 \\ 0 & \lambda_2 - \lambda & \delta_2 & \cdots & 0 & 0 \\ 0 & 0 & \ddots & & & \vdots \\ \vdots & \vdots & & \ddots & \ddots & \vdots \\ 0 & 0 & \cdots & 0 & \lambda_q - \lambda & 0 \\ 2v_1 & 2v_2 & 2v_3 & \cdots & 2v_q & 0 \end{pmatrix}$$

$$= \begin{pmatrix} 0 & \delta_1 & 0 & \cdots & 0 & -1 \\ 0 & \lambda_2 - \lambda & \delta_2 & \cdots & 0 & 0 \\ 0 & 0 & \ddots & & & \vdots \\ \vdots & \vdots & & \ddots & \ddots & \vdots \\ 0 & 0 & \cdots & 0 & \lambda_q - \lambda & 0 \\ 2v_1 & 2v_2 & 2v_3 & \cdots & 2v_q & 0 \end{pmatrix}. \tag{5.37}$$

where the values δ_j are either 0 or 1, depending on whether the corresponding eigenvalues are repeated and defective or not.

5.4.2.1   Non-singular Case

If λ has multiplicity one (i.e., is not repeated), we have $\delta_1 = 0, \forall j = 2,\ldots,q, \lambda_j \neq \lambda$, and:

$$|Df(\vec{u},\lambda)| = \begin{vmatrix} 0 & 0 & 0 & \cdots & 0 & -1 \\ 0 & \lambda_2 - \lambda & \delta_2 & \cdots & 0 & 0 \\ 0 & 0 & \ddots & & & \vdots \\ \vdots & \vdots & & \ddots & \ddots & \vdots \\ 0 & 0 & \cdots & 0 & \lambda_q - \lambda & 0 \\ 2v_1 & 2v_2 & 2v_3 & \cdots & 2v_q & 0 \end{vmatrix}$$

$$= -\begin{vmatrix} 2v_1 & 2v_2 & 2v_3 & \cdots & 2v_q & 0 \\ 0 & \lambda_2 - \lambda & \delta_2 & \cdots & 0 & 0 \\ 0 & 0 & \ddots & & & \vdots \\ \vdots & \vdots & & \ddots & \ddots & \vdots \\ 0 & 0 & \cdots & 0 & \lambda_q - \lambda & 0 \\ 0 & 0 & 0 & \cdots & 0 & -1 \end{vmatrix}$$

$$= 2v_1(\lambda_2 - \lambda)(\lambda_3 - \lambda)\cdots(\lambda_q - \lambda), \tag{5.38}$$

where we have swapped the first and last rows and computed the resulting determinant of a diagonal matrix. The matrix Df can only be singular in this case if $v_1 = 0$, since all the other factors are non-zero. Let us check that, in this case, $v_1 = 1$. The value of v_1, first component of $Q^T \bar{u}$, can be obtained by the scalar product with $(1, 0, \ldots, 0)^T$:

$$v_1 = \left(Q^T \bar{u}\right)^T \begin{pmatrix} 1 \\ 0 \\ \vdots \\ 0 \end{pmatrix} = \bar{u}^T Q Q^{-1} \bar{u} = \bar{u}^T \bar{u} = \|\bar{u}\|^2 = 1, \tag{5.39}$$

where we have used (5.36), and we are assuming that $\bar{u}$ is a solution of (5.30).

5.4.2.2 Singular Cases

Any other situation corresponds to a singular matrix.

If λ is a repeated eigenvalue but not defective, we have (at least) that $\lambda_1 = \lambda_2 = \lambda$ and $\delta_1 = 0$, and, thus, the first two columns of (5.37) are proportional and the matrix is singular.

If λ is a k-times repeated eigenvalue and defective, we have $\lambda_1 = \lambda_2 = \cdots = \lambda_k = \lambda$, $\delta_1 = 1$ and (at least) $\delta_k = 0$: the kth row of (5.30) is null and the matrix is singular.

We conclude that solving (5.30) by Newton's method can give rise to a quadratic convergence. We illustrate this behaviour in the following example.

Example 5.4.1. Let us consider the real matrix

$$A = \begin{pmatrix} 5 & 6 & 2 & -1 & 0 & 2 & 0 & 5 \\ 1 & -7 & 4 & -1 & -4 & 7 & 0 & 0 \\ -1 & 3 & -3 & 2 & 5 & -6 & -1 & -9 \\ 0 & -4 & 6 & 1 & 8 & -4 & -2 & 7 \\ -3 & -8 & 1 & 0 & 4 & -6 & -2 & 6 \\ 9 & 1 & 0 & 0 & -3 & 5 & 2 & 8 \\ -4 & -7 & 2 & 0 & -7 & 1 & 4 & -7 \\ -3 & 8 & 0 & -3 & 8 & -4 & 0 & -1 \end{pmatrix}$$

whose eight different eigenvalues are:

$$\lambda_1 \approx 13.73300591$$
$$\lambda_{2,3} \approx 4.798290215 \pm i2.332369720$$
$$\lambda_{4,5} \approx 1.976179849 \pm i4.084018108$$
$$\lambda_{6,7} \approx -2.504187237 \pm i5.499620471$$
$$\lambda_8 \approx -14.27357156$$

Using Newton's method as explained above, with initial vector $\vec{x}_0 = (0,1,0,0,0,0,0,0)^{\mathrm{T}}$ and initial variable $\eta_0 = r(\vec{x}_0)$, there is convergence to λ_8 and to the corresponding eigenvector

$$\vec{u}_8 \approx \begin{pmatrix} 0.0240048333465025 \\ 0.421094552544371 \\ -0.548915987830902 \\ 0.350443298496940 \\ 0.420175364352729 \\ 0.162848930126086 \\ 0.235672785561606 \\ -0.373329754947481 \end{pmatrix}$$

up to machine-ε double precision in 14 iterations and with almost the same precision in just 10 iterations.

On the other hand, if we use the iterative method (5.6) with $\tau = 0.05$ we achieve convergence at a much slower rate: we obtain after 34 iterations the same precision as with the quadratic method in just 9 steps. Then why use the linear method at all? The problem with Newton's method is that convergence is only guaranteed if the initial seed is *close enough* to the solution. Bounds can be given for this distance but they are not trivial to compute and some variations of Newton's method are usually considered in order to overcome this and other drawbacks [28]. In this case, for instance, the quadratic method does not converge to any solution if we start with a different initial vector among those of the canonical basis. In fact, it is not easy in general to find a priori a suitable initial vector, whereas, for instance in this example, the linear method gives the same kind of convergence for all eight vectors in the canonical basis.

An interesting possibility to enhance the convergence rate is to combine both methods, starting with the linear one and, once we are confident to be near a solution, switch to the quadratic. In Fig. 5.8 we plot the relative errors. We have switched to the quadratic method after 5 iterations of the linear one. A similar result is obtained when computing λ_1 using matrix $-A$ in the linear method: we represent that in Fig. 5.9. The quadratic method is not presented since we have not achieved convergence with it to λ_1 and $\vec{u}_1$ unless properly "seeded" by the linear method.

In order to implement the switch from the linear to the quadratic method, we have compared the results at three consecutive steps and changed when the differences were sufficiently small.

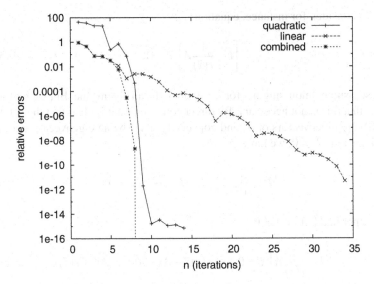

Fig. 5.8 Example 5.4.1. Convergence to λ_8 and $\vec{u}_8$: linear, quadratic and combined methods

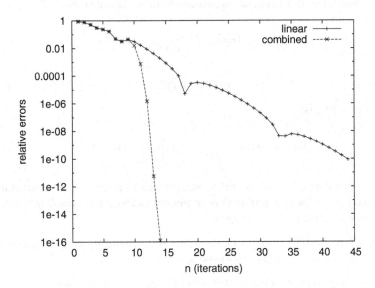

Fig. 5.9 Example 5.4.1. Convergence to λ_1 and $\vec{u}_1$: linear and combined methods

5.4.3   An Alternative Quadratic Method

Besides the previous method, we can explore other possibilities that lead to a quadratic rate of convergence.

Let us consider, for instance, equation

$$\left[I - \frac{1}{r(\vec{x})} A \right] \vec{x} = \vec{0}. \tag{5.40}$$

We must assume A non-singular for $\lambda = 0$ not to be an eigenvalue. But we will see in practice that this is not necessary. By direct computation it is clear that the solutions of (5.40) are eigenvectors of A, and conversely: let $\vec{u}$ be an eigenvector associated with eigenvalue $\lambda \neq 0$, we have

$$\left[I - \frac{1}{r(\vec{u})} A \right] \vec{u} = \left[I - \frac{1}{\lambda} A \right] \vec{u} = \vec{0}. \tag{5.41}$$

On the other hand, if $r(\vec{x}) \neq 0$,

$$\left[I - \frac{1}{r(\vec{x})} A \right] \vec{x} = \vec{0} \Longrightarrow r(\vec{x})\vec{x} - A\vec{x} = \vec{0} \Longrightarrow A\vec{x} = r(\vec{x})\vec{x} \tag{5.42}$$

which is equivalent to $\vec{x}$ being an eigenvector with eigenvalue $r(\vec{x})$.

Let us define

$$G(\vec{x}) = I - \frac{1}{r(\vec{x})} A, \tag{5.43}$$

and

$$\vec{g}(\vec{x}) = G(\vec{x})\vec{x}. \tag{5.44}$$

The Jacobian of $\vec{g}$ is:

$$Dg(\vec{x}) = G(\vec{x}) - \frac{1}{r(\vec{x})} AP(\vec{x}) \left[G(\vec{x}) + G(\vec{x})^{\mathrm{T}} \right], \tag{5.45}$$

where we recall that $P(\vec{x})$ is the orthogonal projector on span$\{\vec{x}\}$. The structure of this method is similar to that of the linear method and we have a result that reminds somewhat one in the previous chapter:

Theorem 5.1. *Basic properties of the alternative quadratic method: let A be a matrix, g, Dg and $\mathscr{P}$ as defined previously. We have:*

1. *$\vec{u}$ is an eigenvector of A associated with an eigenvalue $\lambda \neq 0$, if and only if, $\vec{u}$ is an eigenvector of $Dg(\vec{u})$ with eigenvalue 0.*
2. *Let $\vec{u}$ be an eigenvector of A associated with eigenvalue $\lambda \neq 0$ and let $\vec{w}$ be another eigenvector of A, linearly independent from $\vec{u}$, and associated with eigenvalue μ (which can be equal to λ). Then:*

$$\left[I - P(\vec{u}) \right] \vec{w} \quad \text{is an eigenvector of } Dg(\vec{u}) \text{ with eigenvalue} \quad 1 - \frac{\mu}{\lambda}.$$

Proof. 1.) $\Longrightarrow$) We start from: $A\vec{u} = \lambda\vec{u}, \lambda \neq 0$.

$$Dg(\vec{u})\vec{u} = G(\vec{u})\vec{u} - \frac{1}{r(\vec{u})}AP(\vec{u})\left[G(\vec{u}) + G(\vec{u})^{\mathrm{T}}\right]\vec{u}$$

$$= -\frac{1}{\lambda}A\frac{\vec{u}\vec{u}^{\mathrm{T}}}{\|\vec{u}\|^2}\left[G(\vec{u})\right]^{\mathrm{T}}\vec{u} = -\frac{1}{\lambda}A\frac{\vec{u}}{\|\vec{u}\|^2}\left[G(\vec{u})\vec{u}\right]^{\mathrm{T}}\vec{u}$$

$$= \vec{0}.$$

$\Longleftarrow$) We start from: $\vec{u}$ eigenvector of $Dg(\vec{u})$ with eigenvalue zero.

$$Dg(\vec{u})\vec{u} = \vec{0}, \vec{u} \neq \vec{0} \iff G(\vec{u})\vec{u} - \frac{1}{r(\vec{u})}AP(\vec{u})\left[G(\vec{u}) + G(\vec{u})^{\mathrm{T}}\right]\vec{u} = \vec{0}$$

$$\iff \vec{u} - \frac{1}{r(\vec{u})}A\vec{u} - \frac{1}{r(\vec{u})}A\frac{\vec{u}\vec{u}^{\mathrm{T}}}{\|\vec{u}\|^2}\left[G(\vec{u}) + G(\vec{u})^{\mathrm{T}}\right]\vec{u} = \vec{0}$$

$$\iff \vec{u} = \frac{1}{r(\vec{u})}A\vec{u} + \frac{1}{r(\vec{u})\|\vec{u}\|^2}A\vec{u}\,\vec{u}^{\mathrm{T}}\left[G(\vec{u}) + G(\vec{u})^{\mathrm{T}}\right]\vec{u}$$

and if we define the scalar function $k(\vec{u}) \equiv \vec{u}^{\mathrm{T}}G(\vec{u})\vec{u}$, we have

$$\iff \vec{u} = \frac{1}{r(\vec{u})}A\vec{u} + \frac{2k(\vec{u})}{r(\vec{u})\|\vec{u}\|^2}A\vec{u}$$

$$\iff \vec{u} = \frac{\|\vec{u}\|^2 + 2k(\vec{u})}{r(\vec{u})\|\vec{u}\|^2}A\vec{u}$$

$$\iff A\vec{u} = \lambda\vec{u}, \quad \lambda \equiv \frac{r(\vec{u})\|\vec{u}\|^2}{\|\vec{u}\|^2 + 2k(\vec{u})} \neq 0.$$

2.)

$$Dg(\vec{u})\left[I - P(\vec{u})\right]\vec{w} = Dg(\vec{u})\vec{w} - Dg(\vec{u})\vec{u}\frac{\vec{u}^{\mathrm{T}}\vec{w}}{\|\vec{u}\|^2} = Dg(\vec{u})\vec{w}.$$

On the other hand,

$$Dg(\vec{u})\vec{w} = G(\vec{u})\vec{w} - \frac{1}{r(\vec{u})}AP(\vec{u})\left[G(\vec{u}) + G(\vec{u})^{\mathrm{T}}\right]\vec{w}$$

$$= \left(1 - \frac{\mu}{\lambda}\right)\vec{w} - \frac{1}{\lambda}A\frac{\vec{u}\vec{u}^{\mathrm{T}}}{\|\vec{u}\|^2}\left[G(\vec{u}) + G(\vec{u})^{\mathrm{T}}\right]\vec{w}$$

$$= \left(1 - \frac{\mu}{\lambda}\right)\vec{w} - \left(\vec{u}^{\mathrm{T}}\left[G(\vec{u}) + G(\vec{u})^{\mathrm{T}}\right]\vec{w}\right)\frac{\vec{u}}{\|\vec{u}\|^2}$$

$$= \left(1 - \frac{\mu}{\lambda}\right)\vec{w} - \frac{\vec{u}^{\mathrm{T}}G(\vec{u})\vec{w}}{\|\vec{u}\|^2}\vec{u}$$

$$= \left(1 - \frac{\mu}{\lambda}\right)\vec{w} - \left(1 - \frac{\mu}{\lambda}\right)\frac{\vec{u}^{\mathrm{T}}\vec{w}}{\|\vec{u}\|^2}\vec{u}$$

$$= \left(1 - \frac{\mu}{\lambda}\right)\left[I - P(\vec{u})\right]\vec{w}.$$

□

With all of the above, the resulting Newton's method is given by the iterative process:

$$\vec{x}_{n+1} = \vec{x}_n - \left[Dg(\vec{x}_n)\right]^{-1}\vec{g}(\vec{x}_n) = \left(I - \left[Dg(\vec{x}_n)\right]^{-1}G(\vec{x}_n)\right)\vec{x}_n. \qquad (5.46)$$

In this method, Dg is singular at the solution, but in practice the convergence is, generally, quadratic. We understand that this singularity corresponds to the fact that, even when normalized, the solution is never unique since given a solution $\vec{u}$, $-\vec{u}$ is also another one. But we expect that whenever the eigenvalue is simple or even not defective, we should have quadratic convergence: thanks to the second point of Theorem 5.1, the eigenvectors of $Dg(\vec{u})$ are all orthogonal to $\vec{u}$ and correspond, thus, to independent directions in space. To prove this we should check the Jacobian of Newton's method itself, a task that is not trivial, but it is much simpler to numerically check the convergence. We do this in the following examples.

Example 5.4.2. We use the method on the 5×5 matrices:

Case 1:

$$A_1 = Q\Lambda_1 Q^{-1}, \qquad \Lambda_1 = \begin{pmatrix} -1 & 0 & 0 & 0 & 0 \\ 0 & 1 & 1 & 0 & 0 \\ 0 & 0 & 1 & 0 & 0 \\ 0 & 0 & 0 & 2 & 0 \\ 0 & 0 & 0 & 0 & 3 \end{pmatrix}.$$

Case 2:

$$A_2 = Q\Lambda_2 Q^{-1}, \qquad \Lambda_2 = \begin{pmatrix} -1 & 0 & 0 & 0 & 0 \\ 0 & 1 & 1 & 0 & 0 \\ 0 & 0 & 1 & 0 & 0 \\ 0 & 0 & 0 & 2 & 0 \\ 0 & 0 & 0 & 0 & -3 \end{pmatrix}.$$

Case 3:

$$A_3 = Q\Lambda_3 Q^{-1}, \qquad \Lambda_3 = \begin{pmatrix} -1 & 0 & 0 & 0 & 0 \\ 0 & 1 & 0 & 0 & 0 \\ 0 & 0 & 1 & 0 & 0 \\ 0 & 0 & 0 & 2 & 0 \\ 0 & 0 & 0 & 0 & -3 \end{pmatrix}.$$

We use the same matrix Q for all three cases:

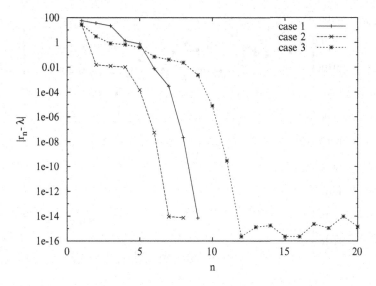

Fig. 5.10 Alternative quadratic method. Example 5.4.2

$$Q = \begin{pmatrix} 1 & 0 & 1 & 1 & 0 \\ 1 & 0 & 0 & 1 & 1 \\ 0 & 1 & -2 & 0 & 1 \\ -1 & 1 & 0 & 1 & -1 \\ 3 & 0 & -1 & -2 & 2 \end{pmatrix},$$

which, thus, fixes the basis for the Jordan's canonical forms.

The convergence is as follows: case 1 towards $\lambda = 3$, case 2 towards $\lambda = 2$ and case 3 towards $\lambda = 1$. In the latter, the solution does not converge to a fixed vector but wanders inside the corresponding eigenspace of dimension 2. The convergence may depend on the initial vector chosen: for all three cases we have started from $(1, -1, 1, -1, 1)^{\mathrm{T}}$. We represent this in Fig. 5.10.

Example 5.4.3. We seek the eigenvalues and eigenvectors of $Q\Lambda Q^{-1}$, with Q as in the previous example, and

$$\Lambda = \begin{pmatrix} -1 & 0 & 0 & 0 & 0 \\ 0 & 0 & 0 & 0 & 0 \\ 0 & 0 & 1 & 0 & 0 \\ 0 & 0 & 0 & 2 & 0 \\ 0 & 0 & 0 & 0 & -3 \end{pmatrix}.$$

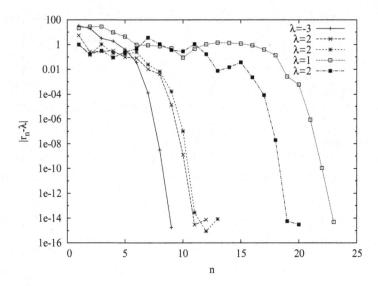

Fig. 5.11 Alternative quadratic method. Example 5.4.3

We choose different initial vectors:

$$\begin{pmatrix} 0 \\ 1 \\ 1 \\ 1 \\ 1 \end{pmatrix}, \begin{pmatrix} 1 \\ 0 \\ 1 \\ 1 \\ 1 \end{pmatrix}, \begin{pmatrix} 1 \\ 1 \\ 0 \\ 1 \\ 1 \end{pmatrix}, \begin{pmatrix} 1 \\ 1 \\ 1 \\ 0 \\ 1 \end{pmatrix}, \begin{pmatrix} 1 \\ 1 \\ 1 \\ 1 \\ 0 \end{pmatrix}.$$

The idea is to sample somewhat the independent directions of the space. We may use the canonical basis, as well. The results are shown in Fig. 5.11.

We do not have convergence towards eigenvalues 0 nor -1. The presence of a zero eigenvalue in A does not imply singularities in the actual computations.

As we have seen in the previous example, some eigenspaces appear to be more elusive than others. We may keep track of the eigenvectors obtained in order to avoid repetitions. As we did in the case of the previous quadratic method, the idea is to combine this with the linear method. We illustrate it through the following example.

Example 5.4.4. Let us apply this second quadratic method to the same matrix as in Example 5.4.1. We start with different initial vectors $\vec{x}_0$, and get convergence to the two only real eigenvalues:
The convergence rates can be seen in Fig. 5.12.

We see that the convergence is quadratic, but only when it really starts moving! Two different behaviours can be observed on the graphic: in the first one (cases 1 and 3) the quadratic convergence is delayed by a "transitory regime" during which

Case	$\vec{x}_0$	Converges to
1	$(1,0,0,0,0,0,0,0)^T$	λ_{min}
2	$(0,1,0,0,0,0,1,0)^T$	λ_{min}
3	$(1,1,0,0,0,0,0,0)^T$	λ_{max}
4	$(0,0,0,0,1,0,0,1)^T$	λ_{max}

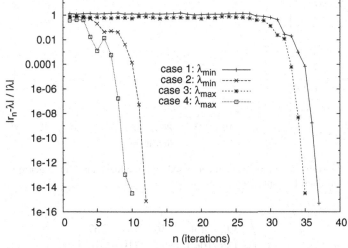

Fig. 5.12 Alternative quadratic method: convergence from different initial vectors towards the two real eigenvalues. Example 5.4.4

the method seems to converge linearly; in the second (cases 2 and 4) the quadratic convergence starts immediately. These two behaviours can be seen also in Fig. 5.11, although in a less dramatic way.

If we compare the results of this method with those of the previous quadratic method in Example 5.4.1, we see that, on the one hand, this alternative method seems to depend less on the initial vector, achieves convergence towards both real eigenvalues, but, on the other, the convergence is not really quadratic during the whole of the process. Once again, coupling the method with the linear one can help avoid this problem, as we see in Fig. 5.13.

Example 5.4.5. We use the alternative quadratic method combined with the linear one to compute the extreme eigenvalues necessary to obtain the optimized values of parameters τ and α for the damped method of Chap. 3. We consider the problem presented in Example 3.1.2, on Page 32: we want the minimum and maximum eigenvalues of matrix $M = A^T A$, with A given by (3.14). We have:

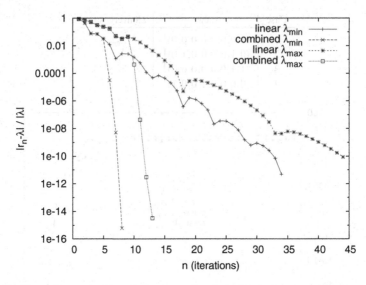

Fig. 5.13 Alternative quadratic method composed with the linear one: convergence towards the two real eigenvalues. Example 5.4.4

$$M = \begin{pmatrix} 1395 & 436 & -36 & 561 & 367 \\ 436 & 1060 & -574 & 519 & -76 \\ -36 & -574 & 1427 & -600 & -315 \\ 561 & 519 & -600 & 1208 & 509 \\ 367 & -76 & -315 & 509 & 1209 \end{pmatrix},$$

and after 32 iterations in the linear regime and 3 in the quadratic one, we obtain:

$$\lambda_{min} \approx 340.8954610153626.$$

Similarly, after 20 iterations in the linear regime and 4 in the quadratic one, we obtain:

$$\lambda_{max} \approx 2855.9655413241339.$$

5.5 Exercises

5.1 In your favourite programming language, implement the numerical method (5.6) and the extension to compute the intermediate eigenvalues using (4.42). Also, implement the combination with both quadratic methods that correspond to applying Newton's method to either (5.30) or (5.44).

5.2 Implement both Power Methods and combine them with the quadratic methods.

5.3 Find, numerically, all the eigenvalues and eigenspaces of matrix A as defined in Exercise 4.4 and check, numerically, the asymptotic behaviour of the method.

5.4 Find, numerically, all the eigenvalues and eigenspaces of matrices

$$
A_1 = \begin{pmatrix} 2 & -12 & 12 & 12 \\ -5 & -1 & -5 & 5 \\ -1 & 7 & -7 & -7 \\ -3 & -5 & 1 & 9 \end{pmatrix}, \quad A_2 = \begin{pmatrix} 8 & 20 & -4 & -20 \\ 3 & 1 & 3 & -3 \\ -5 & -13 & 3 & 13 \\ 5 & 7 & 1 & -9 \end{pmatrix}.
$$

5.5 Compute estimations of $\mu_\pm$ for the Examples and Exercises of Chap. 3.

5.6 We recall that the condition number of a matrix A in the matrix 2-norm is (see Exercise 2.9):

$$
\kappa_2(A) = \|A\|_2 \, \|A^{-1}\|_2 = \frac{\sigma_{max}}{\sigma_{min}}.
$$

Use the numerical method for eigenvectors to compute the condition number of the matrices of the linear systems in Chaps. 2 and 3.

5.7 Find, numerically, the eigenvalues and eigenspaces of matrix A in Exercise 2.9. Choose some value for q, not too big, and take $a = -2$, $b = 1$: compare the eigenvalues you obtain with the analytical expression (2.51).

5.8 Find, numerically, all the eigenvalues and eigenspaces of matrix A in Exercises 4.5 and 4.6. Take advantage of the fact that in both cases A^2 is known.

5.9 Repeat Exercises 4.7–4.9, but using now numerical simulations instead of analytical computations.

5.10 Find, numerically, all the eigenvalues and eigenspaces of matrices J_1 and J_2 of Exercises 4.10–4.13.

5.11 Find, numerically, all the eigenvalues and eigenspaces of the singular matrix (5.21).

5.12 Use the IPMS with a convenient seed to check whether a given matrix has 0 as eigenvalue (and is then singular). Apply it to the matrices of Examples 2.5.1 and 3.3.1, and to (5.21).

5.13 The roots of a polynomial are the eigenvalues of its companion matrix [14]. Matrix M_1 (see Sect. 5.2.2) has characteristic polynomial:

$$
\lambda^5 - 6\lambda^4 + 10\lambda^3 - 11\lambda + 6,
$$

and the corresponding companion matrix is that of Exercise 2.11, b). Find, numerically, all the eigenvectors and eigenvalues of the companion matrix: they are those of M_1. Make use of the sparse character of the matrix to optimize your code.

To find the eigenvalues with intermediate real part, you may use the following property: for a general companion matrix

$$C_q = \begin{pmatrix} 0 & 0 & 0 & 0 & \cdots & -a_0 \\ 1 & 0 & 0 & 0 & \cdots & -a_1 \\ 0 & 1 & 0 & 0 & \cdots & -a_2 \\ 0 & 0 & 1 & 0 & \cdots & -a_3 \\ \vdots & \vdots & \vdots & \ddots & \ddots & \vdots \\ 0 & 0 & 0 & \cdots & 1 & -a_{q-1} \end{pmatrix},$$

we have:

$$C_q^2 = \begin{pmatrix} 0 & 0 & 0 & 0 & \cdots & -a_0 & a_0 a_{q-1} \\ 0 & 0 & 0 & 0 & \cdots & -a_0 & -a_0 + a_1 a_{q-1} \\ 1 & 0 & 0 & 0 & \cdots & -a_1 & -a_1 + a_2 a_{q-1} \\ 0 & 1 & 0 & 0 & \cdots & -a_2 & -a_3 + a_3 a_{q-1} \\ 0 & 0 & 1 & 0 & \cdots & -a_3 & -a_4 + a_4 a_{q-1} \\ \vdots & \vdots & \vdots & \ddots & \ddots & \vdots \\ 0 & 0 & 0 & \cdots & 1 & -a_{q-1} & -a_{q-2} + a_0 a_{q-1} \end{pmatrix}$$

as can be seen from direct multiplication.

5.14 Repeat the previous exercise: for any matrix Λ in Examples 5.4.2 or 5.4.3, build the characteristic polynomial and the companion matrix and find, numerically, the eigenvalues and eigenvectors.

Chapter 6
Linear Programming

6.1 Introduction

We may consider the pinball machine as a mechanical device to visualize the minimization of the potential energy of a ball moving in a bounded inclined plane. The potential energy, in this case, is a linear function of the space coordinates. On the other hand, the boundaries of the inclined plane region, where the ball is moving, are represented by a set of inequalities which define the convex region where the motion is possible. The inequalities are linear if the boundary is made of linear segments, while they are nonlinear if the boundary is a piecewise combination of other kinds of curves.

The *Primal Problem* in linear programming can be written as follows [1,9,10,12]:

$$\max Z, \; Z = \vec{c}^{\mathrm{T}}\vec{x},$$
$$A\vec{x} \leq \vec{b},$$
$$\vec{x} \geq \vec{0}, \tag{6.1}$$

where $\vec{x} \in \mathbb{R}^q$ is the vector with the q independent variables, A is the matrix of dimensions $m \times q$ describing the m linear constraints, $\vec{c} \in \mathbb{R}^q$ is the vector of the coefficients of the (scalar) *objective function* Z and $\vec{b} \in \mathbb{R}^m$ is the vector containing the right-hand side values of the m constraints. The somewhat puzzling notation $A\vec{x} \leq \vec{b}$, widely used by authors in this field, must be understood as an array of inequalities, one for each component of the vector $\vec{b}$, just as $\vec{x} \geq \vec{0}$ should be. We will consider, unless otherwise indicated, the case of *positive coefficients* in the objective function: $\vec{c} \geq \vec{0}$ (using the above-mentioned notation). We assume the standard requirement that the region defined by the constraints, where the maximization of the linear function is to be studied, is convex. This is a most important condition that should be kept mind in all that follows.

L. Vázquez and S. Jiménez, *Newtonian Nonlinear Dynamics for Complex Linear and Optimization Problems*, Nonlinear Systems and Complexity 4, DOI 10.1007/978-1-4614-5912-5_6, © Springer Science+Business Media New York 2013

The basic standard method to solve the Primal Problem is the Simplex Algorithm of Dantzig [3, 9, 10]. Other methods, iterative for instance, have been developed, among which we have Interior Point methods [6].

To compute the solution of the linear programming problem (6.1), we propose an iterative process that can be understood as a path-following method [33]. This method, new to the best of our knowledge, is based on the analysis of the motion of a Newtonian particle in a constant force field [13]. In this case, the force field is either attractive or repulsive. Due to the conservation of mechanical energy, the motion of the particle is such that the potential energy tends to its minimum value. If we define the potential energy as $U = -\vec{c}^{\mathrm{T}}\vec{x}$, then its extreme value will be identified with the maximum of the objective function. The motion is uniformly accelerated and its analytical expression is known:

$$\frac{\mathrm{d}^2\vec{x}}{\mathrm{d}t^2} = \vec{c}, \tag{6.2}$$

where $\vec{x}(t)$, is the q-dimensional displacement of a unit mass under the constant $\vec{c}$ field. The total conserved energy of motion is:

$$E = \frac{1}{2}\left(\frac{\mathrm{d}\vec{x}}{\mathrm{d}t}\right)^2 - \vec{c}^{\mathrm{T}}\vec{x}. \tag{6.3}$$

The solution to the equation of motion (6.2) is given by

$$\vec{x}(t) = \vec{x}(0) + t\frac{\mathrm{d}\vec{x}}{\mathrm{d}t}(0) + \frac{t^2}{2}\vec{c}, \tag{6.4}$$

where $\vec{x}(0)$, $\frac{\mathrm{d}\vec{x}}{\mathrm{d}t}(0)$ are the initial position and velocity, respectively. According to (6.3) the motion of the particle is such that the potential energy $U = -\vec{c}^{\mathrm{T}}\vec{x}$ decreases continuously.

Comparing the mechanical problem with the Primal Problem, we can establish that:

$$\min U = \min(-\vec{c}^{\mathrm{T}}\vec{x}) = -\max(\vec{c}^{\mathrm{T}}\vec{x}) = -\max Z. \tag{6.5}$$

Thus, if we consider the Newtonian particle defined above moving in the constant field $\vec{c}$, starting at rest from the origin (null initial conditions, both for position and velocity) together with the constraints $A\vec{x} \leq \vec{b}$ and $\vec{x} \geq \vec{0}$, the particle will stop at a location such that its potential energy is a local minimum compatible with the given constraints and also at a local maximum of the objective function Z. In principle, it is not assured that starting with null initial conditions (zero initial energy $E = 0$) we shall reach the global minimum value of the objective function under the constraints. But the process can be repeated starting with lower initial values for the energy. This can be achieved, for instance, considering as initial conditions a null velocity $\vec{x}_t(0) = 0$ and an initial position $\vec{x}(0)$ at the centre point of the segment whose extreme points are the position of the local minimum obtained before and the origin. This process stops whenever two consecutive "iterations" give the same local minimum, in the confidence that the minimum obtained is the global one.

In Sect. 6.2, as a means of introduction, we present the trivial case of one variable. Section 6.3 will present several examples for the two-variable case, which analyse the procedure according to the different values of the components of the constant vectors $\vec{b}$ and $\vec{c}$. In Sect. 6.4, rather than presenting the q-variables case in all generality we shall illustrate the general procedure by means of a 4-variable example. Finally in Sect. 6.5 we present some challenges suggested by this method.

6.2 Examples with One Independent Variable

We have, essentially, only two different cases, given by:

$$\begin{aligned} \max Z, \ Z &= ax, \\ x &\leq b, \\ x &\geq 0, \end{aligned} \tag{6.6}$$

and by

$$\begin{aligned} \min Z, \ Z &= ax, \\ x &\geq b, \\ x &\geq 0. \end{aligned} \tag{6.7}$$

We assume $a > 0$. Obviously, the solution to both problems is straightforward. We can associate the solution with the motion of one Newtonian particle starting from $x(0) = 0$ according to the time parametrization $x(t) = \frac{1}{2}at^2$. In both cases, we have to consider the motion until that time at which the particle reaches the boundary, while staying inside of the feasible region. For the maximization problem, the solution is $x = b$, while for the minimization problem it is just zero. Accordingly, the solutions are $Z = ab$ for the maximum case while $Z = 0$ for the minimum one.

These two very simple problems illustrate the basic idea for the proposed process: just as the ball rolling inside the pinball machine, the particle moves freely until it reaches the boundary. At that point, as we will see in the cases with more variables, the motion can stop or continue along the boundary. An important feature is that in some cases we may be able to solve the problem analytically and give the solution parametrized by time. In those cases where this is not possible, or even as an alternate tool for those for which it is, we may compute numerically the solution.

6.3   Examples with Two Independent Variables

In this section we present a set of examples for which graphical visualizations are available. This will give a clearer insight into how our mechanical approach works.

For the sake of clarity some parameters are fixed, despite which the examples still illustrate the general case

Example 6.3.1.

$$\max Z, \ Z = ax + by,$$
$$x + 2y \le 10,$$
$$2x + 3y \le 16,$$
$$2x + y \le 12,$$
$$x, y \ge 0, \tag{6.8}$$

with $a, b \ge 0$. This corresponds to $q = 2, m = 3$:

$$\vec{x} = \begin{pmatrix} x \\ y \end{pmatrix}, \quad \vec{c} = \begin{pmatrix} a \\ b \end{pmatrix}, \quad \vec{b} = \begin{pmatrix} 10 \\ 16 \\ 12 \end{pmatrix}, \quad A = \begin{pmatrix} 1 & 2 \\ 2 & 3 \\ 2 & 1 \end{pmatrix}.$$

The feasible region for this problem is shown in Fig. 6.1 and the optimal solution will occur at one of the extreme points: $(0,0)$, $(0,5)$, $(2,4)$, $(5,2)$ and $(6,0)$. The values of the objective function Z at these points are given in Table 6.1. If we assume that $a \ne 0$ we can define the ratio $\gamma = b/a$: in this way we can give the value of the objective function as depending on just one parameter. We have listed, in the second column of Table 6.1, these values for each extreme point.

Let us apply, now, our mechanical approach to solve problem (6.8). Let us consider the two-dimensional motion of a Newtonian particle with unit mass starting from the origin $(0,0)$ with zero velocity under the constant field force $\vec{c} = (a,b)^{\mathrm{T}}$.

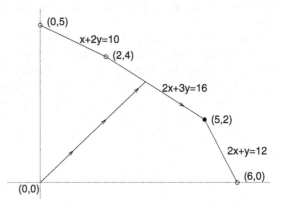

Fig. 6.1 Feasible region for Example 6.3.1. The *arrows* show the visualization of the mechanical iterative approach for $a = 1, b = 1$

Table 6.1 Example 6.3.1:
Values of the objective
function Z at the extreme
points, $\gamma = b/a$

Extreme points	$Z = ax + by$	Z/a
$(0,0)$	0	0
$(0,5)$	$5b$	5γ
$(2,4)$	$2a+4b$	$2+4\gamma$
$(5,2)$	$5a+2b$	$5+2\gamma$
$(6,0)$	$6a$	6

Table 6.2 Example 6.3.1:
Times to reach the linear
constrains

Linear constrains	Time (T^2)	$T/\sqrt{a}$
$x+2y=10$	$\dfrac{20}{a+2b}$	$\sqrt{\dfrac{20}{1+2\gamma}}$
$2x+3y=16$	$\dfrac{32}{2a+3b}$	$\sqrt{\dfrac{32}{2+3\gamma}}$
$2x+y=12$	$\dfrac{24}{2a+b}$	$\sqrt{\dfrac{24}{2+\gamma}}$

The trajectory of the particle is given by the time parametrization:

$$x(t) = \frac{1}{2}at^2, \quad y(t) = \frac{1}{2}bt^2. \tag{6.9}$$

Substituting (6.9) into the inequalities $A\vec{x} \le \vec{b}$, we compute the values of the different times, $T_1, T_2 \ldots, T_m$, at which the particle would reach each one of the straight lines that correspond to the boundaries of the feasible region. These times are the values when the inequalities are satisfied as equalities and they are listed in Table 6.2. In Fig. 6.2 we represent these times as a function of γ.

Given a fixed value of γ we select the minimum of all the Ts: after that time the particle has reached the corresponding boundary and the associated equality holds from that moment on. This allows us to eliminate one of the independent variables and continue with an optimization problem with a reduced number of variables. Taking the minimum time we guarantee that the particle is inside the feasible region of the optimization problem. When performing the reduction of variables in the present case, only one independent variable remains and the result is obtained in a straightforward way.

As a practical application of the method, let us consider the following cases:

1. $a = 1$, $b = 1$. In this case there is an optimal solution. According to Table 6.1, the maximum value of the objective function Z is 7 at the extremal point $(5,2)$. Using our mechanical approach and from Table 6.2, we obtain the minimal value of time $T_2 = \sqrt{32/5}$ that corresponds to reaching the boundary given by the equality $2x + 3y = 16$. We then use this relation to eliminate variable x in the optimization problem, getting:

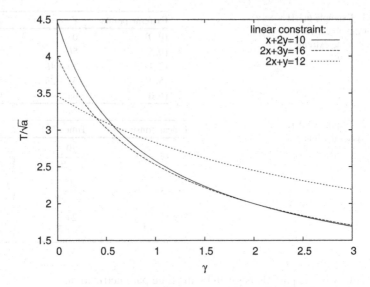

Fig. 6.2 Example 6.3.1. Times to reach the linear constraints as a function of γ

$$\max Z, \ Z = 8 - \frac{3}{2}y + y = 8 - \frac{1}{2}y,$$

$$y \leq 4,$$

$$2 \leq y,$$

$$y \leq \frac{16}{3},$$

$$y \geq 0. \tag{6.10}$$

It is clear that we obtain the maximum value of Z whenever y is minimum: this corresponds to $y = 2$ and, thus, we have $\max Z = 7$ at the point $(5, 2)$ in agreement with the graphical estimation given in Fig. 6.1.

2. $a = 1$, $b = 2$. In this case there is an infinite number of optimal solutions since the graph of the objective function is parallel to the straight line associated with one of the constraints. Substituting the values of a and b in Table 6.1, we see that the objective function has the same maximum value at the two extremum points $(0, 5)$ and $(2, 4)$. In the mechanical approach, we have the minimal value of time $T_1 = T_2 = 4$. This corresponds to the fact that the particle has reached the vertex $(2, 4)$, located on *both* boundaries, at time $T = 4$. In that case, we should take into account both possible simplifications and determine which boundary the particle will follow from that time on.

When we consider $x + 2y = 10$, and eliminate one variable, we get the transformed optimization problem:

$$\max Z, \quad Z = 10,$$
$$y \geq 4,$$
$$y \geq \frac{8}{3},$$
$$y \geq 0,$$
$$5 \geq y. \tag{6.11}$$

The value of Z is now a constant. The new constraints associated with the feasible region imply that $4 \leq y \leq 5$ and $0 \leq x \leq 2$. Thus, *the optimum value is obtained at an infinite number of points* located on the segment determined by the end-points $(0,5)$ and $(2,4)$. When, on the other hand, we select the constraint $2x + 3y = 16$, and we use it to eliminate variable x, we obtain the transformed optimization problem:

$$\max Z, \quad Z = 8 + \frac{1}{2}y,$$
$$y \geq 2,$$
$$y \leq 4,$$
$$y \geq 0, \tag{6.12}$$

which also shows that the optimum value is 10 at the point $(2,4)$. *With this choice we obtain the same optimal value, but there is no information about the non-unicity of the optimal solution location.*

Combining both cases, we see that the particle follows the first boundary we have considered and, in fact, we had already obtained the optimal solution.

Example 6.3.2. Now, let us consider a case where the global optimization value *may not be obtained directly from the analysis of the motion with zero initial conditions*:

$$\max Z, \quad Z = x + \frac{1}{10}y,$$
$$x - 2y \leq 4,$$
$$x - y \leq 5,$$
$$y \leq 2,$$
$$x, y \geq 0, \tag{6.13}$$

which corresponds to

$$\vec{x} = \begin{pmatrix} x \\ y \end{pmatrix}, \quad \vec{c} = \begin{pmatrix} 1 \\ \frac{1}{10} \end{pmatrix}, \quad \vec{b} = \begin{pmatrix} 4 \\ 5 \\ 2 \end{pmatrix}, \quad A = \begin{pmatrix} 1 & -2 \\ 1 & -1 \\ 0 & 1 \end{pmatrix}.$$

Fig. 6.3 Example 6.3.2: feasible region

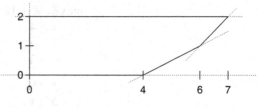

Fig. 6.4 Example 6.3.3. Motion starts outside the feasible region

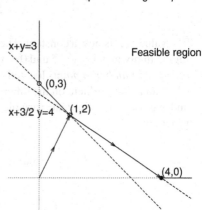

In Fig. 6.3 we represent the feasible region. It is such that the optimal solution will occur at one of the vertex: $(0,0)$, $(0,2)$, $(7,2)$, $(6,1)$ and $(4,0)$.

Let us apply our mechanical approach to solve this problem. If the Newtonian particle starts from the origin $(0,0)$ with zero velocity under the constant field force $\vec{c} = (1, 1/10)^T$, we eventually get the local optimization value $Z = 6.1$ at $(6,1)$. Now, we have to improve the value above in order to get the global optimization value which is $Z = 7.2$ at the point $(7,2)$. To this purpose, we study the motion starting with zero velocity $\vec{x}_t(0) = 0$ at the point $(3, 1/2)$. This point is the mid point of the segment determined by the origin and the local optimization point obtained in the first iteration. Proceeding as before we get the expected global optimization value given above.

Example 6.3.3. Let us consider the following case of minimization of the objective function so as to show how our mechanical approach works in detail:

$$\min Z, \ Z = x + 2y,$$
$$x + \frac{3}{2}y \geq 4,$$
$$x + y \geq 3,$$
$$x, y \geq 0. \tag{6.14}$$

The feasible region (see Fig. 6.4) is such that the optimal solution will occur at one of the extreme points: $(4,0)$, $(1,2)$, and $(0,3)$.

Precisely, the minimum of Z is $Z = 4$ attained at $(4,0)$. Since the optimal value is on the boundary of the feasible region, we can consider two possible motions of our Newtonian particle with unity mass to explore the boundary:

1. Let us consider the starting point located at the origin $(0,0)$ with zero velocity under the constant field force $\vec{c} = (1,2)^{\mathsf{T}}$. By following the trajectory of the particle with the time parametrization we get the time needed by the particle to reach the feasible region, while satisfying all the constraints:

$$x = \frac{t^2}{2}, \quad y = t^2, \tag{6.15}$$

we, thus, obtain the maximum value of time $T_2 = \sqrt{16/5}$, and the corresponding equality $x + \frac{3}{2}y = 4$ which can be used to eliminate the variable x in the optimization problem, getting:

$$\begin{aligned} \min Z, \ &Z = 4 + \frac{1}{2}y, \\ &y \le 2, \\ &y \le \frac{8}{3}, \\ &y \ge 0. \end{aligned} \tag{6.16}$$

Thus we obtain $\min Z = 4$ at the point $(4,0)$ in agreement with the graphical estimation above.

2. A different possibility is to consider a motion starting at an initial point (x_0, y_0) inside the feasible region with null velocity and subject to a constant field force $\vec{c} = (-1, -2)^{\mathsf{T}}$. The time parametrization of the trajectory is in this case given by

$$x = x_0 - \frac{t^2}{2}, \quad y = y_0 - t^2. \tag{6.17}$$

If $x_0 = 4$ and $y_0 = 3$, the particle will arrive to the boundary at time $T = 3/2$ from where we obtain the equality $x = 3/2$ that can be used to eliminate the variable x in the optimization problem, and we get:

$$\begin{aligned} \min Z, \ &Z = 4 + \frac{1}{2}y, \\ &y \le 2, \\ &y \le \frac{8}{3}, \\ &y \ge 0. \end{aligned} \tag{6.18}$$

As before, we get $\min Z = 4$ at the point $(4,0)$. This is shown in Fig. 6.5

Fig. 6.5 Example 6.3.3.
Motion starts inside the
feasible region

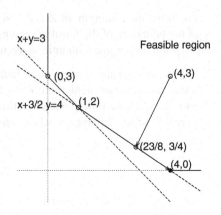

Some Remarks.

1. When there is no solution to the linear programming problem, the so-called case of *infeasibility*, the mechanical solver indicates no solution for the minimization problem for the times of motion associated with every inequality.
2. If the constraint is given by an equality, we can use it to eliminate one of the independent variables and, thus, reduce the complexity of our optimization problem.
3. Without any loss of generality, let us assume that $a \geq 0$ and $b \leq 0$. In order to guarantee the positive constraint on the variables we may start approaching the solution with the motion $x = \frac{1}{2}at^2$ and $y = 0$. Another possible approach, which guarantees the positive constraint, is the motion $x = \frac{1}{2}at^2$ and $y = vt - \frac{1}{2}bt^2$. Since the optimal value is at the boundary of the feasible region, we can explore it with motions inside the feasible region associated with different initial conditions. In this case, we do not have the nice property of the positivity of the coefficients of the objective function but, still, we can use the mechanical motion method to explore and approach the optimal solution.

6.4 General Problem with q Independent Variables

Let us consider the following general problem:

$$\max Z, \ Z = \vec{c}^T \vec{x},$$

$$A\vec{x} \leq \vec{b},$$

$$\vec{x} \geq \vec{0}, \tag{6.19}$$

with $\vec{c} \geq \vec{0}$. Our problem is related to the minimization of the potential energy of a particle moving in the q-dimensional constant force field $\vec{c}$. Let us assume that the particle starts the motion from the origin with zero velocity. The trajectory of the particle is given by

$$\vec{x}(t) = \frac{t^2}{2}\vec{c}.\tag{6.20}$$

By inserting (6.20) in the system $A\vec{x} = \vec{b}$, we obtain q equations, and each one gives us a value T_j, $j = 1,\dots,q$, corresponding to the time when $\vec{x}(T_j)$ satisfies equality j. We then take a value T equal to one of those values such that all the inequalities of (6.19) are satisfied, and one is satisfied as an equality. This enables us to eliminate one of the variables. We repeat the process until there is only one variable left and the problem solved in a straightforward way. At each step the resulting equality limits the feasible region for forward times to a region with one dimension less. Let us illustrate this with some examples, starting with $q = 3$.

Example 6.4.1.

$$\max Z, \quad Z = x + y + 5z,$$
$$x + 2z \le 740,$$
$$7x + 9y + 7z \le 63,$$
$$2x + y + 3z \le 35,$$
$$x, y, z \ge 0.\tag{6.21}$$

By inserting the following expressions in the inequality constraints above

$$x = \frac{1}{2}t^2, \quad y = \frac{1}{2}t^2, \quad z = \frac{5}{2}t^2,\tag{6.22}$$

and solving for the equal sign, we obtain the values

$$T_1 = \sqrt{\frac{1680}{11}} \approx 11.6, \quad T_2 = \sqrt{\frac{42}{17}} \approx 1.6, \quad T_3 = \sqrt{\frac{35}{9}} \approx 2.0.\tag{6.23}$$

From these, the one value such that $\vec{x}(T_j)$ satisfies the three inequalities and guarantees that the particle remains in the feasible region is the smallest time T_2. At that time, the particle has reached the boundary corresponding to the equality $7x + 9y + 7z = 63$ and we use it to eliminate one of the variables, say y. This leaves us with the new optimization problem

$$\max Z, \quad Z = \frac{2}{9}x + \frac{38}{9}z + 7,$$
$$x + 2z \le 740,$$
$$\frac{11}{9}x + \frac{20}{9}z \le 28,$$

$$y \geq 0 \Longrightarrow x + z \leq 9,$$

$$x, z \geq 0. \tag{6.24}$$

We see that the constraint $y \geq 0$ becomes now a new constraint in x and z: $x + z \leq 9$. Following the same ideas as before, we compute the new times and use equality $x + z = 9$ to eliminate one extra variable, say z, and we get the final step of our optimization process

$$\max Z, \;\; Z = -4x + 45,$$

$$x \geq -722,$$

$$x \geq -8,$$

$$x \geq 0. \tag{6.25}$$

From here, and taking into account the equations used to eliminate the variables, we obtain the solution:

$$\max Z = 45,$$

$$x = 0,$$

$$y = 0,$$

$$z = 9. \tag{6.26}$$

We see that in this case we need two iterations of our mechanical approach and the final straightforward computations.

Example 6.4.2. Let us consider the following problem, solved in [25] by the simplex method:

$$\max Z, \;\; Z = 6x + y + 3z - \frac{1}{2}u,$$

$$x + 2z \leq 740,$$

$$2y - 7u \leq 0,$$

$$y - z + 2u \geq \frac{1}{2},$$

$$x + y + z + u = 9,$$

$$x, y, z, u \geq 0. \tag{6.27}$$

Since one of the conditions is an equality, we use it to eliminate one of the independent variables, for instance u, and we obtain an optimization problem with three variables:

$$\max Z, \ Z = \frac{1}{2}x + \frac{1}{2}y + \frac{5}{2}z - \frac{9}{2},$$

$$x + 2z \le 740,$$

$$7x + 9y + 7z \le 63,$$

$$2x + y + 3z \le \frac{35}{2},$$

$$x + y + z \le 9,$$

$$x, y, z \ge 0. \tag{6.28}$$

The time parametrization of the variables corresponds in this case to

$$x = \frac{1}{4}t^2, \quad y = \frac{1}{4}t^2, \quad z = \frac{5}{4}t^2. \tag{6.29}$$

Substituting these values in the four inequality constraints above, and solving for the equality case, we obtain the times

$$T_1 = \sqrt{\frac{2960}{11}} \approx 16.4, \quad T_2 = \sqrt{\frac{84}{17}} \approx 2.2,$$

$$T_3 = \sqrt{\frac{35}{9}} \approx 2.0, \quad T_4 = \sqrt{\frac{36}{7}} \approx 2.3. \tag{6.30}$$

In order to guarantee that the particle remains in the feasible region, the time which satisfies the four inequalities is T_3, thus we consider equality $2x + y + 3z = \frac{35}{2}$ to eliminate variable y, for instance. The new optimization problem is now

$$\max Z, \ Z = \frac{17}{4} - \frac{1}{2}x + z,$$

$$x + 2z \le 740,$$

$$11x + 20z \ge \frac{189}{2},$$

$$x + 2z \ge \frac{17}{2},$$

$$2x + 3z \le \frac{35}{2},$$

$$x, z \ge 0. \tag{6.31}$$

It can be easily seen that the maximal value of Z is obtained for the minimal value of x and the maximal value of z. The position (x,z) of the associated particle can thus be parametrized as:

$$x = 0, \quad z = \frac{1}{2}t^2. \tag{6.32}$$

Following the same process associated with the motion of a particle which starts at the origin and at rest, we insert now this in the four inequality constraints above, in order to guarantee that the associated motion is in the feasible region. This gives us the values:

$$T_1 \lesssim 27.2, \quad T_2 \gtrsim 3.1, \quad T_3 \gtrsim 2.9, \quad T_4 \lesssim 3.4. \tag{6.33}$$

In order to ensure that the particle has arrived to the boundary of the feasible region, we have to select the time $T_2 = 3.1$. Thus, we use the equality $11x + 20z = \frac{189}{2}$ with $x = 0$ to obtain the coordinates of the optimal solution

$$c \max Z = \frac{359}{40} = 8.975,$$

$$x = 0,$$

$$y = \frac{133}{40} = 3.325,$$

$$z = \frac{189}{40} = 4.725,$$

$$u = \frac{19}{20} = 0.95. \tag{6.34}$$

The solution is exactly the one obtained with the simplex method. With our mechanical approach we really need one iteration and the final straightforward calculations, after using the initial equality constraint to reduce the number of variables, to obtain the solution.

6.5   Challenges from the Mechanical Method

From the previous examples, the following general comments pertaining to our mechanical method arise:

1. Being q the number of dimensions of the problem, our mechanical approach implies to carry out $q - 1$ eliminations plus final straightforward calculations, independently of the number m of constraints. For other methods, see, for instance, [20, 21, 27].
2. Following the analysed examples, we can see that our mechanical method gives us a strategy to approach the correct optimal solution. In the general case, it remains open to prove that with our method the obtained solution is optimal and not just a good approximation to the exact solution.

3. The non-uniqueness of the solution for the linear optimization problem is related to different associated mechanical trajectories. In this case, we have to explore such different mechanical trajectories as we did in Example 6.3.1.
4. When some of the coefficients of the objective function are negative, the mechanical method requires the consideration of more general initial conditions for the particle motion, depending on each case.
5. The classical pinball machine mentioned in the Preface represents a good experimental implementation of our mechanical method when the coefficients of the objective function are positive. If some of the coefficients are negative, then it could be visualized through a more complicated pinball machine which should include either electric or magnetic effects on the game ball.

6.6 Exercises

6.1 Find the maximum and the minimum of the function

$$Z = 2x - y,$$

subject to the constraints:

$$3x - y \leq 12,$$
$$x - 2y \geq -6,$$
$$x, y \geq 0.$$

Apply the geometrical, simplex and mechanical methods.

6.2 Compute the maximum and the minimum of the functional

$$Z = x + by$$

with the constraints:

$$x + 2y \geq 2,$$
$$3x + y \geq 4,$$
$$3x + 14y \geq 9,$$
$$x, y \geq 0.$$

6.3 Solve the following minimization problem:

$$\min Z, \quad Z = 3x + 4y,$$
$$2x + 3y \geq 36,$$
$$x + y \geq 14,$$
$$4x + y \geq 16,$$
$$x, y \geq 0.$$

Compare the application of the simplex and mechanical methods.

6.4 Solve the following minimization problem:

$$\min Z, \ Z = 6x + 6y,$$
$$x + 2y \geq 80,$$
$$3x + 2y \geq 160,$$
$$\frac{3}{2}x + 7y \geq 180,$$
$$x, y \geq 0.$$

Compare the application of the simplex and mechanical methods.

6.5 Solve the following minimization problem as a function of a:

$$\min Z, \ Z = 10ax + 8y,$$
$$3x + 2y \geq 30,$$
$$2x + 4y \geq 30,$$
$$x, y \geq 0.$$

6.6 Analyse, in terms of the sign of a, the solution to the problem:

$$\max Z, \ Z = ax + y - 2z,$$
$$x + 2z \leq 10,$$
$$2x + 3y \leq 16,$$
$$2x + y - z \leq 12,$$
$$x, y, z \geq 0.$$

6.7 Solve the minimization problem:

$$\min Z, \ Z = \frac{x}{2} + \frac{y}{2} - 3z,$$
$$4x - 3y - 2z \geq 0,$$
$$3x - 2y + 4z \leq 32,$$
$$2x + 3y - 3z \geq 27,$$
$$x, y, z \geq 0.$$

6.8 Over the same region as in the previous exercise, solve now the maximization problem:

$$\max Z, \ Z = \frac{x}{2} + \frac{y}{2} - 3z.$$

6.9 Solve the minimization problem:

$$\min Z, \ Z = \frac{x}{2} - \frac{y}{4} - 3z,$$

$$-3x - 3y + 2z \leq 6,$$
$$3x + 2y + 2z \leq 90,$$
$$-x + 2y + 2z \leq 50,$$
$$x, y, z \geq 0.$$

6.10 Over the same region as in the previous exercise, solve now the maximization problem:

$$\max Z, \ Z = \frac{x}{2} - \frac{y}{4} - 3z.$$

6.11 Solve the minimization problem:

$$\min Z, \ Z = -2x + y - z,$$

$$-2x - 3y + 2z \leq 4,$$
$$x - y + 2z \leq 10,$$
$$-x + 2y + 2z \leq 25,$$
$$x, y, z \geq 0.$$

6.12 Over the same region as in the previous exercise, solve now the maximization problem:

$$\max Z, \ Z = -2x + y - z.$$

6.13 Solve the minimization problem:

$$\min Z, \ Z = -2x + y - z,$$

$$-3x - 3y + 2z \leq 3,$$
$$x - y + 2z \leq 10,$$
$$-2x + y + 2z \leq 20,$$
$$x, y, z \geq 0.$$

6.14 Over the same region as in the previous exercise, solve now the maximization problem:

$$\max Z, \ Z = Z = -2x + y - z.$$

Chapter 7
Quadratic Programming

7.1 Introduction

In Chap. 6, we associated the minimization of a linear functional with linear constraints to the motion of a Newtonian particle in a constant gravitational field in a bounded region with the frontier made of straight segments. Now, we can extend this mechanical picture to visualize the minimization of a quadratic functional with constraints which can be either linear or nonlinear. Mechanically, the solution is associated with the motion of a Newtonian particle in a quadratic potential with damping and with the associated geometrical constraints. In many cases the analytical estimations are available and we do not need to resort to the numerical simulations. Also, this is the picture in the case of the minimization of a nonlinear functional

In this chapter, we extend the mechanical method developed in Chap. 6 to approach more general optimization problems [6]. In Sects. 7.2 and 7.3 we deal with the case of quadratic objective functions with linear constraints. In Sect. 7.4 some examples of the linear objective function with quadratic constraints are presented. For these optimization problems, at least some preliminary analytical considerations are possible, because the associated motion of the Newtonian particle is either in a constant field or in a harmonic oscillator-like force field. Finally, we consider a general statement about our mechanical approach for the nonlinear programming problem: as for the linear programming case, our mechanical approach represents a strategy to obtain a solution, and, eventually, to obtain the optimal solution running again the iterative process with different initial conditions.

L. Vázquez and S. Jiménez, *Newtonian Nonlinear Dynamics for Complex Linear and Optimization Problems*, Nonlinear Systems and Complexity 4, DOI 10.1007/978-1-4614-5912-5_7, © Springer Science+Business Media New York 2013

7.2 Quadratic Programming

Let us consider the quadratic optimization problem

$$\min Z, \ \ Z = \frac{1}{2}\vec{x}^{\mathrm{T}} M \vec{x} - \vec{c}^{\mathrm{T}} \vec{x},$$

$$A\vec{x} \geq \vec{b},$$

$$\vec{x} \geq \vec{0}, \tag{7.1}$$

where $\vec{x}$ and $\vec{c}$ are vectors of q components, M is a $q \times q$ symmetric real matrix, A is the matrix associated with the m constraints and $\vec{b}$ is the vector containing the right-hand side values of the m constraints. As in the previous chapter, vector inequalities correspond to an array of inequalities, one for each component.

If there were no constraints, and M is positive definite, the minimization problem would amount to solve the linear system $M\vec{x} = \vec{c}$. The mechanical method to solve this unconstrained problem was discussed in Chaps. 2 and 3. In the present general case (7.1), the solution for the constrained system can be obtained following the q-dimensional trajectory of a unit mass Newtonian particle in the potential defined by the objective function Z and in the presence of a linear dissipation. The equation of motion is

$$\frac{d^2\vec{x}}{dt^2} = \vec{c} - M\vec{x} - \alpha\frac{d\vec{x}}{dt}, \tag{7.2}$$

where $\vec{x}(t)$ is the q-dimensional displacement of the particle, and $\alpha > 0$. The energy of such a particle is:

$$E = \frac{1}{2}\left(\frac{d\vec{x}}{dt}\right)^2 - \vec{c}^{\mathrm{T}}\vec{x} + \frac{1}{2}\vec{x}^{\mathrm{T}} M \vec{x}. \tag{7.3}$$

Due to the dissipation, the energy is not preserved and, as we indicated in Chap. 1, we have

$$\frac{dE}{dt} = -\alpha\left(\frac{d\vec{x}}{dt}\right)^2. \tag{7.4}$$

If the constraints in (7.1) are not present, the trajectory defined by (7.2) tends to the solution of the unconstrained problem $M\vec{x} = \vec{c}$, as previously mentioned. We now have a natural algorithm to solve the minimization problem, similar to our procedure for the linear programming: starting from some initial conditions, we follow a trajectory until it reaches the boundary of the feasible region and use the corresponding constraint to reduce our problem. We repeat this process until a solution is obtained. Then, we may check whether this is optimal or not.

In order to obtain the trajectory we may follow two different approaches. The first one is to consider the analytical solution of (7.2). Written as a system, and defining an auxiliary vector $\vec{p} = \dot{\vec{x}}(t)$, in a similar way to what was done in Sect. 2.3 for the discrete case, we have, in stacked and boxed notation:

$$\begin{pmatrix} \vec{x}(t) \\ \vec{p}(t) \end{pmatrix} = e^{Nt} \begin{pmatrix} \vec{x}(0) \\ \vec{p}(0) \end{pmatrix}, \quad N = \left(\begin{array}{c|c} O & I \\ \hline -M & -\alpha I \end{array} \right). \tag{7.5}$$

Evaluating e^{Nt} can be a lengthy process, depending on your favourite method to implement the exponential of a matrix (Jordan canonical form, Fulmer method, etc.) but, in any case, it is simplified here since M is a real symmetric matrix (and thus has a canonical diagonal form) and since the eigenvalues and eigenvectors of M and N are related in the following way:

Theorem 7.1. *Let $\vec{u}$ be an eigenvector of M with eigenvalue μ. Then,* $\begin{pmatrix} \vec{u} \\ \lambda_+\vec{u} \end{pmatrix}$ *and*

$\begin{pmatrix} \vec{u} \\ \lambda_-\vec{u} \end{pmatrix}$ *are two eigenvectors of N associated, respectively, to eigenvalues*

$$\lambda_+ = \frac{-\alpha + \sqrt{\alpha^2 - 4\mu}}{2}, \quad \lambda_- = \frac{-\alpha - \sqrt{\alpha^2 - 4\mu}}{2}.$$

Proof. In close similarity to what was done to obtain (2.31), we have

$$\lambda \text{ is eigenvalue of } N \iff \left| \begin{array}{c|c} -\lambda I & I \\ \hline -M & -(\alpha+\lambda)I \end{array} \right| = 0$$

$$\iff \left| \begin{array}{c|c} O & I \\ \hline -[M+\lambda(\alpha+\lambda)I] & -(\alpha+\lambda)I \end{array} \right| = 0 \iff |M+\lambda(\alpha+\lambda)I| = 0$$

$$\iff -\lambda(\alpha+\lambda) \text{ is eigenvalue of } M.$$

If $\vec{u}$ is now an eigenvector of M with eigenvalue μ, we have

$$-\lambda(\alpha+\lambda) = \mu \iff \lambda_\pm = \frac{-\alpha \pm \sqrt{\alpha^2 - 4\mu}}{2},$$

and by substitution we confirm that the eigenvectors associated with $\lambda_\pm$ are $\begin{pmatrix} \vec{u} \\ \lambda_\pm\vec{u} \end{pmatrix}$, respectively. $\qquad\square$

The second possibility is to numerically integrate equation (7.2) with the dissipative Størmer–Verlet scheme (2.20), for instance, while monitoring at every time step the evolution of the objective function and the linear constraints which define the boundary associated with the problem. The existence of the discrete energy (2.23) and its variation law (2.22) ensure that our numerical solution will evolve towards lower values of the objective function.

Once the solution, computed either analytically or numerically, reaches the boundary of the feasible region, we use the associated local linear boundary equation to eliminate one of the variables as in our mechanical linear programming method of Chap. 6. Then, we repeat the process but with $q - 1$ variables, starting from the corresponding point on that boundary. In the numerical case we could determine it by a cubic interpolation, for instance. On the other hand, since the exact value is not in general relevant, we may use the value given by the simpler linear interpolation.

Remarks. 1. When matrix M is diagonal, the analytical solution is easy to obtain, since the equations for each component of $\vec{x}$ are decoupled, and in general it is not necessary to resort to the numerical simulations.

2. Still, when the matrix M is diagonal and, furthermore, $\vec{c} = \vec{0}$, the problem can also be solved analytically by a different approach. The motion of the particle in the potential $\frac{1}{2}\vec{x}^T M \vec{x}$ is then described by a simple expression: each component is a trigonometric function of time. With these we compute the first crossing of the feasible region boundary in order to eliminate one of the variables and to repeat the process with $n - 1$ variables, and there is no need to consider a dissipation. We illustrate this in Example 7.3.1, below.

3. When M is not diagonal, instead of using either the analytical approach (7.5) or the numerical one (2.20), we could consider a coordinate transformation in order to diagonalize the matrix M and proceed as indicated in the previous remarks. The extra computational cost that this requires may turn this option into an uninteresting one, nevertheless. Also, the possibility of finding analytically the coordinate transformation in the case of five or more dimensions is not guaranteed.

4. It is important to verify whether or not the origin, $\vec{x} = \vec{0}$, lies inside of the feasible region. On the other hand, the origin is always the absolute minimum of the potential function $\frac{1}{2}\vec{x}^T M \vec{x}$ if M is a symmetric and positive definite, real matrix.

5. In the general case, the "recipe" is to start following the trajectory of the associated particle from a point inside the feasible region and from rest. Only, when the origin belongs to the feasible region and the vector $\vec{c}$ is not null, we can start the motion of the associated particle in the origin and at rest.

7.3   Quadratic Objective Function with Linear Constraints

7.3.1   One Dimension

In this case, $Z = \frac{1}{2}ax^2 - cx$, and the constraint can be written in general as $x \geq b$. The problem can be solved analytically and compared with the numerical solution. For the sake of simplicity, let us consider $b = 0$, otherwise we can perform a change of variable and reduce the problem to this one. According to the signs of a and c we

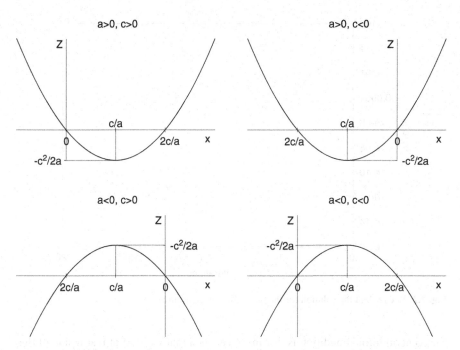

Fig. 7.1 Quadratic objective function

Table 7.1 Minimum value
of the quadratic objective
function

a	c	$\min Z$
$a > 0$	$c > 0$	$-c^2/2a$
$a > 0$	$c < 0$	0
$a < 0$	$c > 0$	unbounded from below
$a < 0$	$c < 0$	local minimum at 0 unbounded from below

have four general cases which are represented in Fig. 7.1 and the solution indicated
in Table 7.1. In the first three cases there is only one solution for the minimization
problem. In the last case $a < 0$, $c < 0$, there is a local minimum at $x = 0$ but Z is
unbounded from below.

Now, let us apply our mechanical method by integrating numerically the
equation. Since we are considering a scalar problem, (2.20) simply becomes (1.15)
and substituting the value of the force for this case, we have:

$$\frac{x_{n+1} - 2x_n + x_{n-1}}{\tau^2} = c - ax_n - \alpha \frac{x_{n+1} - x_{n-1}}{2\tau},$$

$$\forall n, x_n \geq 0. \tag{7.6}$$

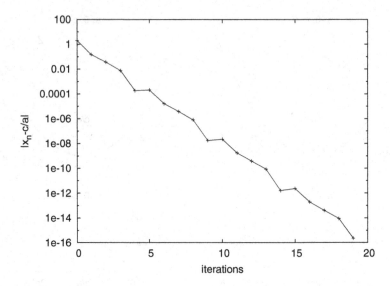

Fig. 7.2 Location of the solution for the quadratic objective function, $a = c = 1$

The general initial condition is, for instance, such that we start at rest inside of the feasible region: $x(0) > 0$, $\frac{dx}{dt}(0) = 0$. We can apply the analysis we developed in Chap. 2 to obtain conditions for efficient values of parameters τ and α. For $a > 0$, we have in this case $\alpha = \sqrt{2a}$, $\tau\alpha < 2$. Since it its interesting to have $\tau\alpha$ as close to 2 as possible, we have chosen $\tau = 1.9/\sqrt{2a}$. In Fig. 7.2 we have represented the absolute error of x_n with respect to the location of the solution in the case $a = c = 1$, starting from $x_0 = 3$.

With the same values of α and τ as above, in the case $a = 1$, $c = -1$, the solution reaches the boundary $x = 0$ after just one iteration independently of the starting value x_0. In this one-dimensional case, this means we have obtained the solution $x = 0$ in just two steps. In the case $a < 0$, $c < 0$, depending on the initial condition, we obtain either 0 as solution, when we reach the boundary $x = 0$, or no solution, with x_n growing in an unbounded manner. This last behaviour is also the one observed in the case $a < 0$, $c > 0$.

7.3.2   Two Dimensions

As mentioned in Remark 3 above, in certain cases we can implement some analytical computations avoiding both the use of the general framework and of the numerical simulations: this is done in the following example.

Fig. 7.3 Feasible region
(*shaded in gray*) for (7.7)

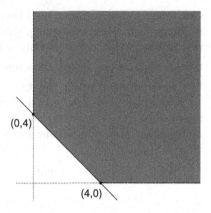

(0,4)

(4,0)

Example 7.3.1.

$$\min Z, \; Z = \frac{1}{2}x^2 + \frac{1}{2}y^2,$$

$$x + y \geq 4,$$

$$x, y \geq 0. \tag{7.7}$$

The feasible region is shown in Fig. 7.3. It does not include the origin $(0,0)$, the
point where the objective function Z would have its absolute minimum value if
there were no constraints.

Let us consider the motion of a particle in the two-dimensional potential Z, with
no dissipation. The equations of motion are:

$$\frac{d^2x}{dt^2} = -x,$$

$$\frac{d^2y}{dt^2} = -y,$$

$$x + y \geq 4,$$

$$x, y \geq 0. \tag{7.8}$$

Assuming that the particle is initially at rest at a given point, $(x(0), y(0))$, inside the
feasible region, the time parametrization of the trajectory would be given by

$$x(t) = x(0)\cos(t),$$

$$y(t) = y(0)\cos(t),$$

$$x + y \geq 4,$$

$$x, y \geq 0. \tag{7.9}$$

Due to the parametrization, both boundaries $x = 0$ and $y = 0$ would be reached simultaneously. This is something impossible according to the form of the feasible region. The particle can, thus, only reach the boundary $x + y = 4$ and it does so at time $T = \arccos\left(\frac{4}{x(0)+y(0)}\right)$. This time T does not need to be computed: we only need to know that it exists, and it is so in this case since inside the feasible region we have $x(0) + y(0) \geq 4$. Once the particle is on the boundary, for $t \geq T$, we use the equality $x(t) + y(t) = 4$ to eliminate one of the two variables and we reduce the optimization problem to the one-dimensional case:

$$\min Z, \ \ Z = \frac{1}{2}x^2 + \frac{1}{2}(4 - x)^2,$$
$$4 - x \geq 0,$$
$$x \geq 0, \tag{7.10}$$

equivalent to

$$\min Z, \ \ Z = x^2 - 4x + 16,$$
$$4 \geq x \geq 0, \tag{7.11}$$

which can then be solved. The optimal solution is obtained at point $(x, y) = (2, 2)$ and the minimum value is $Z = 4$.

If we apply the general framework to this problem, we have

$$M = \begin{pmatrix} 1 & 0 \\ 0 & 1 \end{pmatrix}, \quad \mu_1 = \mu_2 = 1, \quad N = \begin{pmatrix} 0 & 0 & 1 & 0 \\ 0 & 0 & 0 & 1 \\ -1 & 0 & -\alpha & 0 \\ 0 & -1 & 0 & -\alpha \end{pmatrix}. \tag{7.12}$$

If $\alpha \neq 2$, then $\lambda_+ \neq \lambda_-$ and:

$$e^{Nt} = \frac{1}{\lambda_+ - \lambda_-}\mathcal{E},$$

$$\mathcal{E} = \begin{pmatrix} \lambda_+ e^{\lambda_- t} - \lambda_- e^{\lambda_+ t} & 0 & e^{\lambda_+ t} - e^{\lambda_- t} & 0 \\ 0 & \lambda_+ e^{\lambda_- t} - \lambda_- e^{\lambda_+ t} & 0 & e^{\lambda_+ t} - e^{\lambda_- t} \\ e^{\lambda_- t} - e^{\lambda_+ t} & 0 & \lambda_+ e^{\lambda_+ t} - \lambda_- e^{\lambda_- t} & 0 \\ 0 & e^{\lambda_- t} - e^{\lambda_+ t} & 0 & \lambda_+ e^{\lambda_+ t} - \lambda_- e^{\lambda_- t} \end{pmatrix},$$

$$\tag{7.13}$$

where $\lambda_{\pm}$ are the values defined in Theorem 7.1. On the other hand, for the particular case $\alpha = 2$, we have $\lambda_+ = \lambda_- = -1$, and:

Fig. 7.4 Numerical
simulations for problem (7.7)

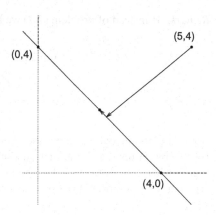

$$e^{Nt} = \begin{pmatrix} (1+t)e^{-t} & 0 & te^{-t} & 0 \\ 0 & (1+t)e^{-t} & 0 & te^{-t} \\ -te^{-t} & 0 & (1-t)e^{-t} & 0 \\ 0 & -te^{-t} & 0 & (1-t)e^{-t} \end{pmatrix}. \tag{7.14}$$

If we remain within this simpler case, and consider that initially the particle is at
rest, we have:

$$x(t) = (1+t)e^{-t}x(0), \quad y(t) = (1+t)e^{-t}y(0), \tag{7.15}$$

which are always positive. As before, the only boundary that can be reached is
$x+y=4$, and this is done at a time T such that

$$(1+T)e^{-T} = \frac{4}{x(0)+y(0)}. \tag{7.16}$$

Once again, we do not need to compute this value. We then simplify the problem
considering $y = 4 - x$ and we obtain the new problem (7.11), as before.

It is clear that this analysis, though straightforward, is somewhat lengthy and
a better choice could be to simulate numerically the problem using (2.20). In
this simple case, since the equations for x_n and y_n are uncoupled, this would be
equivalent to simulate two independent scalar equations using for each of them
the numerical scheme (7.6), with a choice of parameters similar to what was
considered in the one-dimensional case: $\alpha = \sqrt{2}$, $\tau = 1.9/\sqrt{2}$. Starting from a point
$(x(0),y(0))$ inside the feasible region, the particle reaches the boundary $x+y=4$
in just one iteration. We then simplify the problem and obtain (7.11). We represent
this in Fig. 7.4, starting from $(x(0),y(0)) = (5,4)$.

It is clearly seen from this example how numerical simulations may be the most
effective and simplest way to solve this problem.

Remark. If instead of problem (7.7) we had the problem

$$\min Z, \ \ Z = \frac{1}{2}x^2 + \frac{1}{2}y^2,$$

$$x + y \leq 4,$$

$$x, y \geq 0, \tag{7.17}$$

where the first inequality is reversed, the feasible region is the complement of that for (7.7) with respect to the first quadrant. The origin belongs now to the feasible region where the function Z has the absolute minimum and, thus, by mechanical arguments the solution of our optimal problem is $x = 0$, $y = 0$.

7.4   Linear Objective Function with Quadratic Constraints

The computational strategy is the same than the previous one developed for the case of linear objective functions and linear constraints. The motion of an associated particle is parametrized as before, since this part of the problem has not changed. The difference now is that the boundaries are not linear but, much in the same way as before, they are used to reduce the number of variables.

7.4.1   One Dimension

This case is very simple but illustrates some aspects that we find in larger dimensions. We apply the same mechanical considerations as before:

$$\min Z, \ \ Z = cx,$$

$$ax^2 \geq b,$$

$$x \geq 0. \tag{7.18}$$

If we consider $a, b, c > 0$, the minimum value of Z is $c\sqrt{b/a}$ at $x = \sqrt{b/a}$, as it is shown if Fig. 7.5. The origin $x = 0$ is not inside the feasible region. Following what was done in the previous chapter, we can associate the solution of the problem whic the motion of a particle starting at the origin, at rest, with constant acceleration c until it reaches the boundary of the feasible region. The free motion corresponds to $x(t) = \frac{1}{2}ct^2$ and the time to reach the boundary is T such that $ax(T)^2 = b$. As a consequence, the optimal value is $Z = c\sqrt{b/a}$. A similar analysis for other signs of the parameters a, b, c may be done in a straightforward way.

Fig. 7.5 Graphic
representation for problem
(7.18)

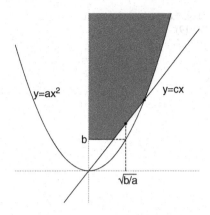

7.4.2   Two Dimensions

Let us consider the case studied in Page 463 of [3] in order to see how our approach works differently:

$$\min Z, \quad Z = -2x - y,$$

$$x^2 + y^2 \le 25,$$

$$x^2 - y^2 \le 7,$$

$$x \le 5, \; y \le 10,$$

$$x, y \ge 0. \tag{7.19}$$

In Fig. 7.6 we plot the feasible region for this problem. We remark that it is not a convex region. The origin $(x,y) = (0,0)$ is on its border. Identifying the function Z with a potential, a particle, initially at the origin and at rest, will move in such field force towards the absolute minimum of Z according to the time parametrization: $x = t^2$, $y = \frac{1}{2}t^2$. Leaving the origin under the influence of the potential, the particle would reach the boundary $x^2 + y^2 = 25$ at time $T_1 = (20)^{1/4}$, and would reach $x^2 - y^2 = 7$ at time $T_2 = (28/3)^{1/4}$. We may use the hyperbolic boundary to eliminate one of the variables, say x, and, thus, obtain the final minimization problem:

$$\min Z, \quad Z = -2(y^2 + 7)^{1/2} - y,$$

$$0 \le y \le 3. \tag{7.20}$$

It is clear now that the minimum occurs at the maximal value of y, and the optimal solution is $Z = -11$, obtained at $(x,y) = (4,3)$.

Fig. 7.6 Feasible region for (7.19)

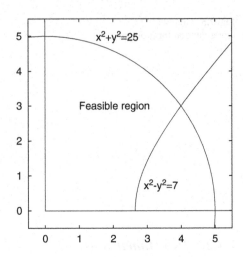

7.5 Extension to Nonlinear Programming

The approach developed for the linear and quadratic programming can be extended to the nonlinear programming. The general nonlinear objective function can be considered as associated with the potential energy of a Newtonian particle moving in a region defined by the constraints of the problem. In this case, there is no general analytical solution for the unconstrained associated problem, contrarily to that of linear programming and many cases of the quadratic programming. Thus, we have to implement a suitable numerical algorithm as the introduced for the quadratic programming (2.20). A presentation of this kind of numerical schemes for this case can be found in [18]. We describe this through the following example.

Example 7.5.1. Let be the nonlinear problem

$$\min Z, \ Z = x^4 + y^4,$$
$$2x + y \geq 4,$$
$$2x + 5y \geq 10,$$
$$x, y \geq 0. \tag{7.21}$$

If we consider Z as the potential, the equations of the motion for a particle moving under the corresponding force, and a linear dissipative term, are

$$\frac{d^2x}{dt^2} = -4x^3 - \alpha \frac{dx}{dt},$$
$$\frac{d^2y}{dt^2} = -4y^3 - \alpha \frac{dy}{dt}. \tag{7.22}$$

Although uncoupled, these nonlinear differential equations cannot be solved in terms of elementary functions. We use a numerical scheme to simulate them. Each equation can be solved independently. We may consider, as we have done before, the use of the Størmer–Verlet scheme (1.15). But in this case of a cubic force, there is no discrete energy that follows an exact dissipation law and the numerical solutions may show an unwanted behaviour. In order to avoid this, we may use the Strauss–Vázquez scheme (1.16).

The expression of the numerical methods in this case are:

$$\frac{x_{n+1} - 2x_n + x_{n-1}}{\tau^2} = -4x_n^3 - \alpha \frac{x_{n+1} - x_{n-1}}{2\tau}, \tag{7.23}$$

for Størmer–Verlet, and for Strauss–Vázquez:

$$\frac{x_{n+1} - 2x_n + x_{n-1}}{\tau^2} = -\left(x_{n+1}^3 + x_{n+1}^2 x_{n-1} + x_{n+1} x_{n-1}^2 + x_{n-1}^3\right)$$

$$-\alpha \frac{x_{n+1} - x_{n-1}}{2\tau}. \tag{7.24}$$

We only give the expressions for variable x, since they are just the same for y, with the appropriate substitutions. The analysis we did before for parameters α and τ is no longer valid in this nonlinear case, since the corresponding eigenvalues of the Jacobian matrix that govern the linear behaviour and stability near the fixed points are not constant but depend on x_n. We may, nevertheless, use the values of the linear problem as a reference.

For instance, choosing $\alpha = \sqrt{2}$ and $\tau = 1.9/\alpha$, we reach a boundary in one iteration (two steps). In Fig. 7.7, we represent the values obtained for different initial data inside the feasible region, computed with the Størmer–Verlet method. The Strauss–Vázquez method gives similar quantitative values. As we see, either boundary can be reached.

Once on the boundary we express one variable in terms of the other. We have two cases:

1. $y = 4 - 2x$: the new problem is

$$\min Z, \ Z = x^4 + (2x - 4)^4,$$

$$x \leq \frac{5}{4},$$

$$x \geq 0. \tag{7.25}$$

This problem is easily solved. It can be done analytically, since the potential Z is a decreasing function in the interval $[0, 5/4]$. It can also be done numerically, and the solution is: $x = 5/4$, $y = 3/2$, $\min Z = 1921/256$.

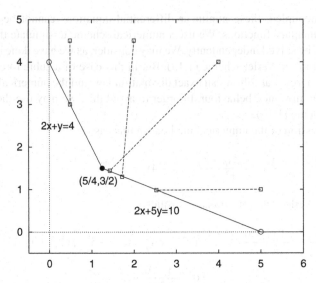

Fig. 7.7 Feasible region and numerical simulations for (7.21). The initial points are inside the feasibility region, the points at the next iteration step are on the boundary

2. $y = 2 - \dfrac{2}{5}x$: the reduced problem is now

$$\min Z, \; Z = x^4 + \left(\frac{2}{5}x - 2\right)^4,$$

$$x \ge \frac{5}{4},$$

$$x \le 5. \tag{7.26}$$

In this case, the potential is an increasing function in the interval $[5/4, 5]$ and the solution is, as before, $x = 5/4$, $y = 3/2$, $\min Z = 1921/256$, which we could also have computed numerically.

Example 7.5.2. This is an example with nonlinear objective function and some nonlinear constraints. We want to solve the minimization problem:

$$\min Z, \; Z = x^4 - y^4,$$

$$x^2 + y^2 \le 25,$$

$$x^2 - y^2 \le 7,$$

$$-(x - 3)^2 + y \ge 0,$$

$$x, y \ge 0. \tag{7.27}$$

The feasible region is represented in Fig. 7.8.

Fig. 7.8 Feasible region and numerical simulations for Example 7.5.2

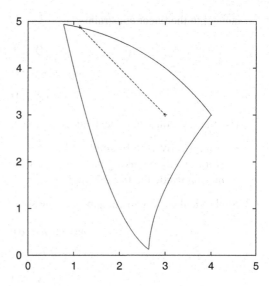

We proceed as we did for Example 7.5.1 and perform a numerical simulation with the Størmer–Verlet method. We have started from the point $(3,3)$ with null velocity and reached the boundary $x^2 + y^2 = 25$ in one iteration. We substitute the value for x^2, for instance, and have the new problem:

$$\min Z, \quad Z = (5^2 - y^2)^2 - y^4 = 25(25 - 2y^2),$$
$$y^2 + 6\sqrt{25 - y^2} + y \leq 34,$$
$$y \geq 3. \tag{7.28}$$

Since Z is now a decreasing function of y, the minimum is obtained at the maximal value of y. This corresponds thus to

$$y^2 + 6\sqrt{25 - y^2} + y = 34, \quad y \geq 3. \tag{7.29}$$

We have solved the equation numerically and obtained $y \approx 4.939167463$, $Z \approx -594.7687615$ and $x \approx 0.7775762165$.

7.6 Exercises

7.1 Solve the following minimization problem

$$\min Z, \quad Z = x^2 + \frac{1}{4}y^2,$$

subject to the linear constraints:

$$x+2y \geq 1,$$
$$x-y \geq 0,$$
$$x,y \geq 0. \tag{7.30}$$

Use the two consistent mechanical approaches:

- Analytically as in Example 7.3.1.
- Numerically by taking into account the damping, Eq. (7.2), and starting the motion inside the feasible region.

7.2 Study via the mechanical approach the solutions to the following system

$$\max Z, \ Z = ax+by,$$
$$x+2y \leq 10,$$
$$x^2 - y^2 \leq 1,$$
$$x,y \geq 0,$$

where a and b have no definite sign.

7.3 Study the solution to the following minimization problem

$$\min Z, \ Z = x^2 + ay^2 + bxy,$$
$$3x+2y \geq 9,$$
$$2x+4y \geq 4,$$
$$x,y \geq 0,$$

where a and b have no definite sign.

7.4 Obtain the solution of the minimization problem

$$\min Z, \ Z = (x-3)^2 + 2y^2,$$
$$(1-x)^5 - y \geq a,$$
$$x,y \geq 0,$$

where $a \geq 0$.

7.5 With the mechanical approach, study the minimum of the following problem considered in [4]:

$$\min Z, \ Z = 2x^2 + 2y^2 + z^2 + 2xy + 2xz + 9 - 8x - 6y - 4z,$$
$$x+y+z \leq 3,$$
$$x,y,z \geq 0.$$

7.6 Study the minimum of the following problem

$$\min Z, \ Z = x^8 + y^8,$$
$$2x + y \geq 4,$$
$$2x + 5y \geq 10,$$
$$x, y \geq 0.$$

7.7 Study the minimum of the following problem

$$\min Z, \ Z = x^4 + y^7,$$
$$2x + y \geq 4,$$
$$2x + 5y \geq 10,$$
$$x, y \geq 0.$$

7.8 Solve the quadratic problem:

$$\min Z, \ Z = x^2 + y^2 - 3z^2,$$
$$4x - 3y - 2z \geq 0,$$
$$3x - 2y + 4z \leq 32,$$
$$2x + 3y - 3z \geq 27,$$
$$x, y, z \geq 0.$$

7.9 Solve the quadratic problem:

$$\min Z, \ Z = x^2 - y^2 - 2z^2 + yz,$$
$$-3x - 3y + 2z \leq 6,$$
$$3x + 2y + 2z \leq 90,$$
$$-x + 2y + 2z \leq 50,$$
$$x, y, z \geq 0.$$

7.10 Solve the quadratic problem with quadratic constraints:

$$\min Z, \ Z = x^2 - y,$$
$$x^2 + y^2 \leq 25,$$
$$x^2 - y^2 \leq 7,$$
$$x \leq 5, \ y \leq 10,$$
$$x, y \geq 0.$$

References

1. D.R. Anderson, D.J. Sweeney, T.A. Williams, *Linear Programming for Decision Making* (West Publishing, New York, 1974)
2. P.M. Anselone, L.B. Rall, The solution of characteristic value-vector problems by Newton's method. Numer. Math. **11**, 38–45 (1968)
3. M. Avriel, *Nonlinear Programming. Analysis and Methods* (Dover Publications, Mineola, 2003)
4. E.M.L. Beale, Numerical Methods in *Nonlinear Programming*, ed. by J. Abadie (North Holland Publishing, Amsterdam, 1967)
5. J.T. Betts, *Practical Methods for Optimal Control and Estimation Using Nonlinear Programming*, 2nd edn. SIAM's Advances in Design and Control (2010)
6. J.F. Bonnans, J.Ch. Gilbert, C. Lemaréchal, C.A. Sagastizábal, *Numerical Optimization: Theoretical and Practical Aspects* (Springer, New York, 2006)
7. F. Chatelin, *Eigenvalues of Matrices* (Wiley, Chichester, 1995)
8. K.A. Cliffe, T.J. Garratt, A. Spence, Eigenvalues of block matrices arising from problems in Fluid Mechanics. SIAM J. Matrix Anal. Appl. **15**(4), 1310–1318 (1994).
9. R. Cottle, E. Johnson, R. Wets, George B. Dantzig (1914–2005). Not. AMS **54**(3), 344–369 (2007)
10. G.B. Dantzig, *Linear Programming and Extensions* (Princeton University Press, Princeton, 1963)
11. V.N. Faddeeva, *Computational Methods of Linear Algebra* (Dover Publications, New York, 1959)
12. J. Franklin, *Methods of Mathematical Economics* (Springer, New York, 1980)
13. H. Goldstein, *Classical Mechanics* (Addison-Wesley, Readings, 1981)
14. G.H. Golub, Ch.F. Van Loan, *Matrix Computations*, 2nd edn. (Johns Hopkins, Baltimore, 1989)
15. J. Guckenheimer, P. Holmes, *Nonlinear Oscillations, Dynamical Systems, and Bifurcations of Vectors Fields* (Springer, New York, 1983)
16. E. Hairer, C. Lubich, G. Wanner, *Geometric Numerical Integration*, 2nd edn. (Springer, New York, 2006)
17. S. Jiménez, L. Vázquez, Analysis of a nonlinear Klein-Gordon equation. Appl. Math. Comput. **35**, 61–94 (1990)
18. S. Jiménez, P. Pascual, C. Aguirre, L. Vázquez, A panoramic view of some perturbed nonlinear wave equations. Int. J. Bifurcat. Chaos **14**(1), 1–40 (2004)
19. S. Jiménez, L. Vázquez, A dynamics approach to the computation of eigenvectors of matrices. J. Comput. Math. **23**(6), 657–672 (2005)
20. N. Karmarkar, A new Polynomial-time algorithm for linear programming. Combinatorica **4**(4), 373–395 (1984)

L. Vázquez and S. Jiménez, *Newtonian Nonlinear Dynamics for Complex Linear and Optimization Problems*, Nonlinear Systems and Complexity 4, DOI 10.1007/978-1-4614-5912-5, © Springer Science+Business Media New York 2013

21. L.G. Khachiyan, A polynomial Algorithm in Linear Programming. Dokl. Akad. Nauk SSSR, **244**(S), 1093–1096 (1979), translated in *Soviet Mathematics Doklady* **20**(1), 191–194 (1979)
22. V.V. Konotop, L. Vázquez, *Nonlinear Random Waves* (World Scientific, Singapore, 1994). See also references [379], [403], [326], [404] and [191], therein.
23. M.C. Navarro, H. Herrero, A.M. Mancho, A. Wathen, Efficient solution of a generalized eigenvalue problem arising in a thermoconvective instability. Comm. Comput. Phys. **3**(2), 308–329 (2008)
24. L. Perko, *Differential Equations and Dynamical Systems*, 3rd edn. (Springer, New York, 2001)
25. W.H. Press, S.A. Teukolsky, W.T. Vetterling, B.P. Flannery, J.G.P. Barnes, *Numerical Recipes in C. The Art of Scientific Computing*, 2nd edn. (Cambridge University Press, Cambridge, 1995)
26. M. Rossignoli, *The Complete Pinball Book: Collecting the Game & Its History* (Schiffer Publishing, Atglen, 2011)
27. F. Santos, A counterexample to the Hirsch conjecture, arXiv:1006.2814 (2010)
28. J. Stoer, R. Burslisch, *Introduction to Numerical Analysis*, 2nd edn. (Springer, New York, 2002)
29. W.A. Strauss, L. Vázquez, Numerical solutions of a nonlinear Klein-Gordon equation. J. Comput. Phys. **28**, 271–278 (1978)
30. J. Todd, The condition number of the finite segment of the Hilbert matrix. Natl. Bur. Stand. Appl. Math. Ser. **39**, 109–116 (1954)
31. L. Vázquez, S. Jiménez, Analysis of a mechanical solver for linear systems of equations. J. Comput. Math. **19**(1), 9–14 (2001)
32. L. Vázquez, J.L. Vázquez-Poletti, A new approach to solve systems of linear equations. J. Comput. Math. **19**(4), 445–448 (2001)
33. L. Vázquez, J.L. Vázquez-Poletti, A mechanical approach for linear programming. http://www.uni-bielefeld.de/ZiF/complexity; *ZiF Preprint* 2001/066 (2001)

Index

L. Vázquez and S. Jiménez, *Newtonian Nonlinear Dynamics for Complex Linear and Optimization Problems*, Nonlinear Systems and Complexity 4, DOI 10.1007/978-1-4614-5912-5, © Springer Science+Business Media New York 2013